M. J. Mochane
and T. C. Mokhena

Handbook of Carbon-Based Filler(s) Reinforced PLA Biocomposites

Fundamentals to Applications

Copyright © 2023 by Nova Science Publishers, Inc.

All rights reserved. No part of this book may be reproduced, stored in a retrieval system or transmitted in any form or by any means: electronic, electrostatic, magnetic, tape, mechanical photocopying, recording or otherwise without the written permission of the Publisher.

We have partnered with Copyright Clearance Center to make it easy for you to obtain permissions to reuse content from this publication. Please visit copyright.com and search by Title, ISBN, or ISSN.

For further questions about using the service on copyright.com, please contact:

	Copyright Clearance Center	
Phone: +1-(978) 750-8400	Fax: +1-(978) 750-4470	E-mail: info@copyright.com

NOTICE TO THE READER

The Publisher has taken reasonable care in the preparation of this book but makes no expressed or implied warranty of any kind and assumes no responsibility for any errors or omissions. No liability is assumed for incidental or consequential damages in connection with or arising out of information contained in this book. The Publisher shall not be liable for any special, consequential, or exemplary damages resulting, in whole or in part, from the readers' use of, or reliance upon, this material. Any parts of this book based on government reports are so indicated and copyright is claimed for those parts to the extent applicable to compilations of such works.

Independent verification should be sought for any data, advice or recommendations contained in this book. In addition, no responsibility is assumed by the Publisher for any injury and/or damage to persons or property arising from any methods, products, instructions, ideas or otherwise contained in this publication.

This publication is designed to provide accurate and authoritative information with regards to the subject matter covered herein. It is sold with the clear understanding that the Publisher is not engaged in rendering legal or any other professional services. If legal or any other expert assistance is required, the services of a competent person should be sought. FROM A DECLARATION OF PARTICIPANTS JOINTLY ADOPTED BY A COMMITTEE OF THE AMERICAN BAR ASSOCIATION AND A COMMITTEE OF PUBLISHERS.

Library of Congress Cataloging-in-Publication Data

ISBN: 979-8-89113-184-2

Published by Nova Science Publishers, Inc. † New York

Contents

Preface vii

Acknowledgements ix

Chapter 1   **Carbon Based Fillers: Synthesis, Structure, and Properties** 1
                 Abstract.. 1
                 1.1. Introduction 1
                 1.2. Carbon-Based Nanoparticles 5
                     1.2.1. Carbon Nanotubes.................................... 5
                     1.2.2. Graphite.. 9
                 1.3. Future Recommendations 14
                 1.4. Conclusion 15
                 References 15

Chapter 2   **PLA Matrix: Synthesis Route, Structure, and Properties** 19
                 Abstract.. 19
                 2.1. Introduction 19
                 2.2. Methods of Obtaining Lactide 25
                 2.3. The Properties of the Lactide 25
                 2.4. Synthesis of PLA 26
                 2.5. Incorporation of Other Groups into PLA Matrix: Efforts to Functionalize PLA 31
                 2.6. Future Recommendations 36
                 2.7. Conclusion 36
                 References 37

Chapter 3   **Preparation and Morphology of Carbon-Based Filler(s) Reinforced PLA** 39
                 Abstract.. 39
                 3.1. Introduction 39

	3.2. Preparation of PLA/CBFs Composites 40
	3.2.1. Melt-Blending 42
	3.2.2. Secondary Melt-Blending 46
	3.2.3. Solution Mixing 52
	3.2.4. In Situ Polymerization 62
	3.3. Conclusion 65
	References 66
Chapter 4	**Thermal Stability of PLA/Carbon Based Filler(s) Composites 73**
	Abstract 73
	4.1. Introduction 73
	4.2. Thermal Properties 74
	4.2.1. CNTs-Based Composites 75
	4.2.2. Graphene-Based PLA Composites 83
	4.2.3. CB-Based Composites 87
	4.3. Hybridization 88
	4.4. Conclusion 88
	References 89
Chapter 5	**Flammability Properties of PLA/Carbon Based Filler(s) Composites 93**
	Abstract 93
	5.1. Introduction 93
	5.2. Flammability Properties: Graphite and Its Derivates/PLA Based Nanocomposites 96
	5.3. Flammability Properties: PLA/Carbon Nanotubes Composites 106
	5.4. Future Recommendations 109
	5.5. Conclusion 109
	References 110
Chapter 6	**Mechanical Properties of the PLA/ Carbon Based Fillers Composites 113**
	Abstract 113
	6.1. Introduction 113
	6.2. Factors Affecting Mechanical Properties of PLA 115
	6.3. Mechanical Properties of PLA-CBF Composites 116
	6.3.1. Solution Mixing 117

	6.3.2. Electrospinning	*124*
	6.3.3. Melt-Compounding	*125*
	6.3.4. Extrusion	*129*
	6.3.5. Three-Dimensional (3D) Printing	*130*
	6.3.6. Melt-Spinning	*135*
	6.4. Hybridization	136
	6.5. Conclusion	139
	References	140
Chapter 7	**Applications of PLA/Carbon Based Fillers Composites**	**147**
	Abstract	147
	7.1. Introduction	147
	7.2. Applications of CBF-PLA Composites	149
	7.2.1. Packaging	*149*
	7.2.2. Healthcare Applications	*152*
	7.2.3. Sensing Applications	*155*
	7.2.4. Oil-Water Separation	*157*
	7.3. Conclusion	158
	References	158
About the Authors		**163**
Index		**165**

Preface

This book extensively explores the effect of various carbon-based fillers on the properties of PLA for advanced applications i.e., packaging, health, sensing, and oil-water separation. Poly(lactic acid) is a multifaceted eco-friendly biopolymer that can be produced from renewable resources such as wheat, sugar, and corn. PLA is utilized as an alternative polymer for replacement of fossil fuel derived polymers, and as a result it has attracted many applications. Nevertheless, as much as PLA is utilized in various applications, there is a need to modify the properties of the PLA in order to widen the applications of the PLA. To improve properties such as mechanical properties, flame retardancy, thermal stability and thermal properties, carbon-based fillers are added into the PLA matrix. Various carbon-based fillers such as carbon nanotubes, graphite and its derivatives have been incorporated in the PLA matrix, as a result giving a balance in terms of properties *viz* mechanical, thermal and flame resistance. The first part of the book is based on the synthesis, structure, and properties of the carbon-based fillers. There has been steady increase in the study of carbon nanomaterials in the last six years. The book further discusses the synthesis, and properties of the PLA, which also showed a steady increase in publications from 2017 to 2022. Finally, the book discusses the properties of PLA/carbon-based nanoparticles for advances applications. The book further highlights the importance of eco-friendly composites as a future composite of interest for both economic and environmental awareness. The book will be useful for scientists, students, engineers, and professionals in the working in the field of biocomposites.

Acknowledgements

The authors would like to acknowledge the following institutions:

- Central University of Technology, Free State (South Africa)
- National research foundation (South Africa)
- Mintek (South Africa)

Chapter 1

Carbon Based Fillers: Synthesis, Structure, and Properties

Abstract

Carbon based fillers (CBFs) have been used globally as reinforcing materials because of their novel properties and promising applications. CBFs are also utilized with various matrices to obtain composite materials with superior electrical, thermal, and mechanical performances. The well-known CBFs fillers are graphene, graphene oxide, carbon nanotubes, fullerenes, expandable and expanded graphite. This chapter discusses various synthesis routes for carbon-based fillers. Methods such as the chemical vapour deposition (CVD), mechanical exfoliation, chemical exfoliation, epitaxial growth, and pyrolysis have been used for the synthesis of carbon based fillers. Furthermore, the chapter discusses the properties of various carbon-based fillers.

Keywords: carbon, carbon-based fillers, synthesis, bottom-up, top-up

1.1. Introduction

Carbon is one of the most common elements in the period table and it is classified as a non-metal (see scheme 1, symbol A). Since ancient times, carbon has been found in various allotropes such as graphite, amorphous carbon, and diamond. Carbon and the component associated with it are estimated to contribute 0.032% of the Earth's crust. Carbon is the most important element in the Periodic Table due to its ability to form single bonds with itself, and as a result is able to withstand chemical attack at room temperature. Furthermore, carbon is an integral part of many compounds, including living cells such as the DNA.

It is well known that bulk quantities of the carbon are present in the form of compounds. Carbon-based nanomaterials (CBN) with carbon as a key member in the family include carbon nanotubes, fullerenes, carbon black, graphite, graphene, graphene oxide, carbon-based quantum dots and diamonds [1, 2]. According to the statistics below (Figure 1.1), the field of chemistry was found to be leading, followed by material science, with the tenth field found to be biochemistry molecular biology. Generally, carbon-based nanomaterials have been studied extensively due to their high surface area, better mechanical strength, and high thermal conductivity etc. [3-10]. For example, there has been a steady increase in the number of publications in relation to the study of the carbon-based nanomaterials. Figure 1.2 illustrates such an increase in the last five years (2017-2022), even though there is no final list of publications for 2022, as the data were generated in the year 2022. Based on the high number of publications (Figure 1.2) for carbon-based nanomaterials, CBN has been utilized in various applications. For example, CBNs (graphite and carbon nanotubes) have been used for (i) enhancing thermal and electrical conductivities of the polymers; (ii) Carbon black has been incorporated into the rubber for automotive applications; and (iii) in biomedical applications (biosensor, cancer therapy, drug delivery).

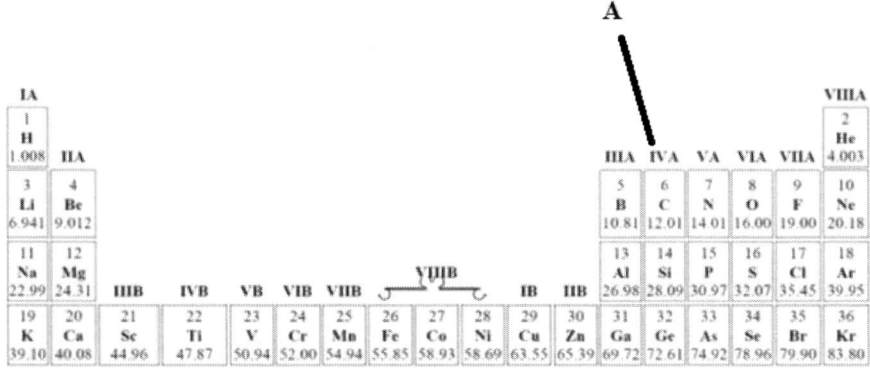

Scheme 1. The first 36 elements in the Periodic Table.

Due to a high interest in the field of carbon based nanoparticles, there has been a few leading countries on the study or field of carbon nanomaterials. The People's Republic of China is currently the leading country, followed by the United States of America (USA), with Poland (Figure 1.3) being the tenth country in terms of the studies on carbon-based nanomaterials. Furthermore, Figure 1.4 illustrates the citations by countries in the study of carbon-based nanomaterials.

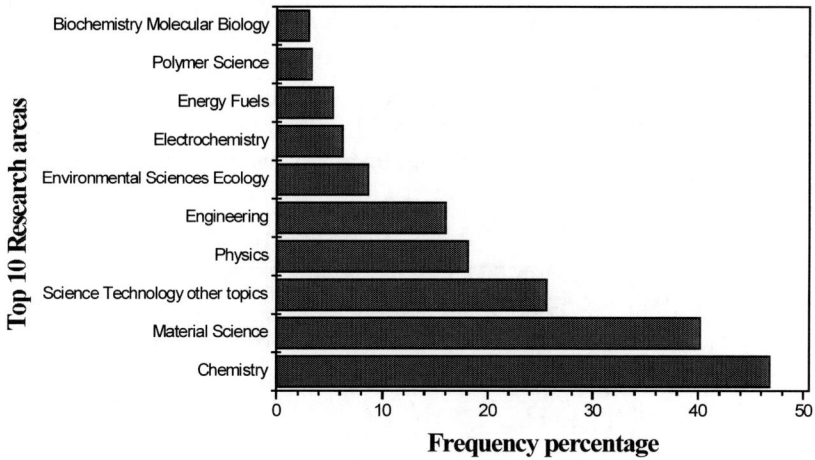

Figure 1.1. Leading fields by typing the keywords: Carbon nanomaterials (Date:12 August 2022).

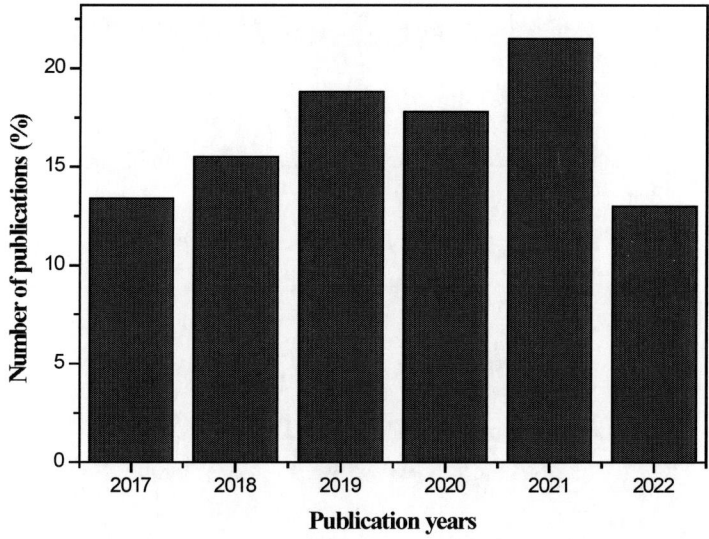

Figure 1.2. Number of publications from 2017-2022, by typing the Carbon nanomaterials (Date: 12 August 2022).

However, as much as the carbon-based nanomaterials are utilized in various applications, the synthesis methods of this nanoparticles play a key role in terms of the resultant structure and properties. Various methods have been studied for the fabrication of the carbon-based nanoparticles, including laser ablation [11], plasma-enhanced chemical vapor deposition (PECVD) [12], arc discharge [13], high-pressure conversion of carbon monoxide (HiPco) [14], micromechanical exfoliation [15], liquid-phase exfoliation (LPE) [16], pulsed laser deposition (PLD) [17], and solvothermal synthesis [18]. This chapter provides an in-depth discussion on the fabrication methods for carbon-based nanoparticles.

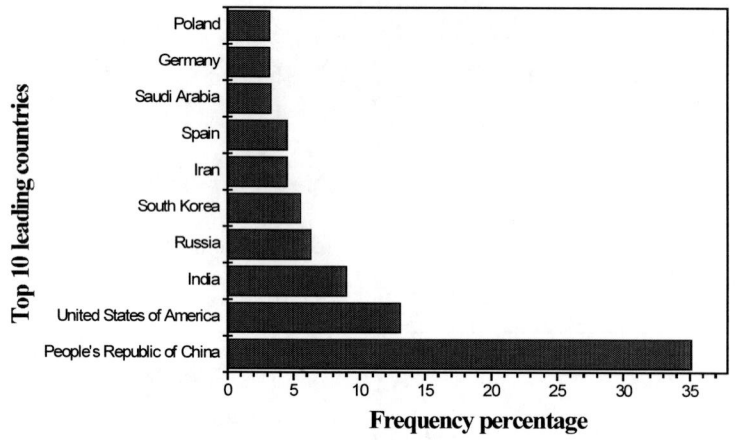

Figure 1.3. Leading countries by typing the keywords: Carbon nanomaterials (Date:12 August 2022).

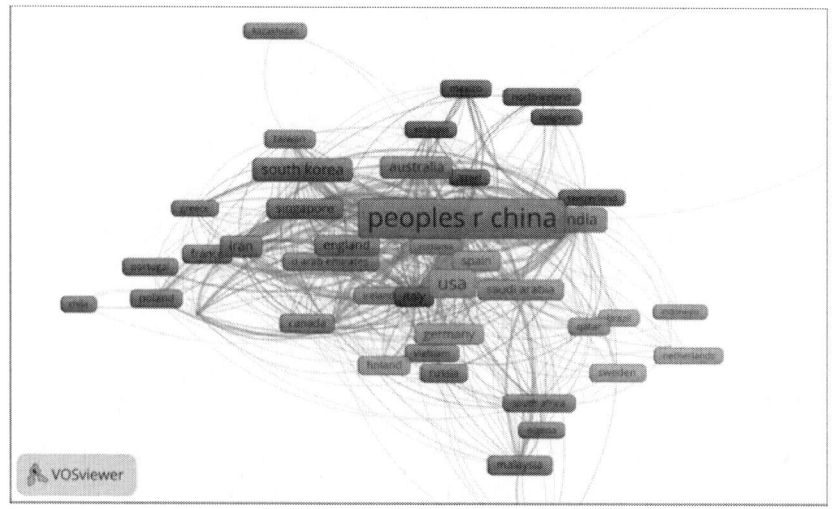

Figure 1.4. Illustration of the citations by country (Date: 12 August 2022).

1.2. Carbon-Based Nanoparticles

This section discusses various synthesis methods for carbon-based nanoparticles such as (i) carbon nanotubes, (ii) graphite, (iii) graphene, (iv) graphene oxide, (v) carbon black, and (vi) fullerenes. Because the synthesis methods affect the morphology, the morphology of the resultant nanoparticles is also discussed and correlated with the properties.

1.2.1. Carbon Nanotubes

Carbon nanotube (multi-walled carbon nanotubes) were discovered by Sumio Iijima in 1991 through the method called the arc-discharge method [19]. The single-walled carbon nanotubes were discovered two years later in 1993 through the discovery of multiwalled carbon nanotubes (MWCNTs) [20]. The single-walled nanotube (SWCNT) has a cylindrical tube fabricated with one layer of the graphene rolled side to side. The multi-walled carbon nanotubes (MWCNTs) consist of concentric cylinders of graphene sheets with various diameters. There is a 0.32 nm to 0.35 nm interplanar distance between the concentric cylinders. It is apparent that between the two types of carbon nanotubes, the MWCNTs are the preferred ones, because they are less expensive to produce than the SWCNTs. Furthermore, there is a special type of multi-walled carbon nanotubes consisting of only two concentric cylinders of graphene. This type of MWCNT is called the double-walled carbon nanotube (DWCNT). It is noted that the DWCNT has similar properties than those of the SWCNT. Table 1.1 summarizes the differences between multi-walled carbon nanotubes and single-walled carbon nanotubes.

Table 1.1. Comparison between MWCNT and SWCNT [21]

MWCNTs	SWCNTs
Multi-layers of graphene	A layer of graphene sheets
May be used in absence of a catalyst	Catalyst is used during synthesis
There is an ease of bulk synthesis	Bulk synthesis is not easy
Highly purified	Poor purity
It cannot be twisted easily	Ease of twisting
Complex structure	Less complex structure; as a result ease of evaluation and characterization

1.2.1.1. Synthesis Route for Carbon Nanotubes and Their Morphology

As explained in the introduction section, various methods have been utilized for the fabrication of the carbon nanotubes (*viz* MWCNT and SWCNT). Cao and co-workers [22] report on the synthesis of the single-walled carbon nanotubes (SWCNT) via the chemical vapor deposition of acetylene (C_2H_2). Since it was elaborated in Table 1.1 that a catalyst is required in most synthesis of the SWCNT, in this case, the catalyst utilized was a mixture of Fe/Mo/Co on an Al_2O_3. The catalyst was chosen because Mo-based catalysts are well known for high-yield production of the synthesized CNTs. Factors such as the content of the catalyst, growth time, and growth temperature were studied in depth in this study. The growth time utilized in this study was 30, 60 and 90 min at temperatures such as 600, 750 and 800°C. It was reported that the density of CNTs grown was found to be lower at lower temperatures (i.e., 600°C) (Figure 1.5a); however, there was a higher density of the CNTs grown at 750 (Figure 1.5b) and 800°C (Figure 1.5c), when compared with the 600°C. One can deduce that active nucleation sites for the development of CNTs are fabricated at higher temperatures. The density of the CNTs was found to be denser at the same temperature for longer growth time (Figure 1.6).

Figure 1.5. SEM images of CNTs grown on a Fe/Mo/Co/ Al_2O_3 at 5/1/1/80 (wt%) for a period of 1 hour: (a) 600°C, b) 750°C and c) 800°C. The flow rates of Ar: $H_2:C_2H_2$ = 420:100: 14 sccm [22].

Chrzanowska et al. [23] report on the synthesis carbon nanotubes by employing the laser ablation method. The effect of laser wavelength on the synthesis and properties of SWCNTs was investigated. The wavelengths employed in this study were 355 and 1064 nm. SEM images revealed CNTs nanoparticles with a diameter of about 1.25 nm. Elsewhere in the literature [24], Yttria catalyst was used for the synthesis of the carbon nanotubes by the arc discharge method. The arc discharge method was used due to its ability to produce both multi-walled carbon nanotubes (MWCNTs) and single-walled carbon nanotubes (SWCNTs). However, the method has been popular in terms of producing the finest SWCNTs [24]. The basic principle of the arc discharge is such that the source material is converted into a vapour via an arch discharge located between the electrodes, with the process followed by condensation, and nucleation, as well as nanoparticle growth. Figure 1.7 shows a typical illustration of the arc discharge experiment.

(a) (b)

Figure 1.6. SEM images of the CNTs growth for period of (a) 30 min and 60 min at 750°C [22].

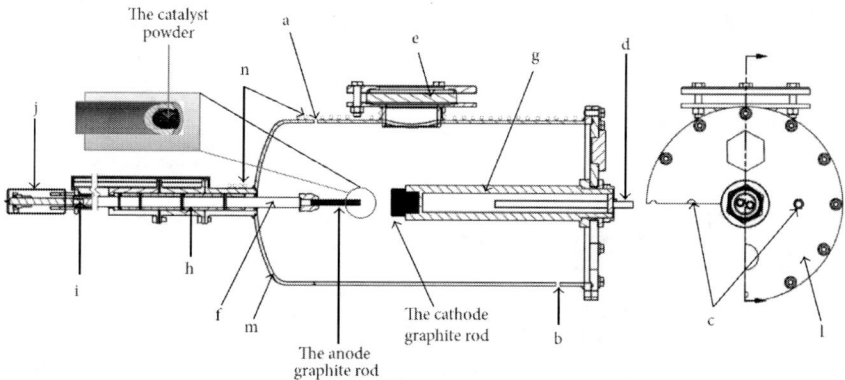

Figure 1.7. A typical illustration of the arc discharge experiment employed by Mohammad et al. [24]. The symbols in Figure 1.8 are noted as (a) a vacuum outlet, (b) gas inlet, (c) indicators of pressure and vacuum gages, d) cathode inlet and outlet coolant water, (e) sealed quartz windows, (f) high-purity copper shaft coating with nickel layer, (g) high-purity copper tube isolated from the chamber, (h) feed through, (i) lead screw, (j) steeper motor, (l) sealed flange cover, (m) stainless steel chamber, and (n) copper pipe for water coolant [24].

There are factors which were found to affect the production of the carbon nanotubes during the process of arch discharge such as the plasma density, plasma sheath thickness, electric field, and the catalyst [24]. To be specific, amongst the factors mentioned above, a catalyst was found to be more effective in affecting the yield of the process significantly. For example, it was reported that the liquid-state metals or their synergy is known to enhance the yield [25]. In the study [24] the production of carbon nanotubes was synthesized with a mixture of an yttrium oxide/nickel/carbon catalyst. The synergy of the catalyst had the following % ratios: 1% Y_2O_3, 4% Ni and 95% of carbon, with the particle sizes of 12 µm, 7.5 µm, and 20 µm, respectively. The SEM images (Figure 1.8 a and b) of the tubes produced in the absence of the catalysts revealed a carbonaceous network with the needles interacting to form bundles, while Figure 1.8(c) and (d) revealed carbonaceous residues with fairly large tube diameters under high magnifications.

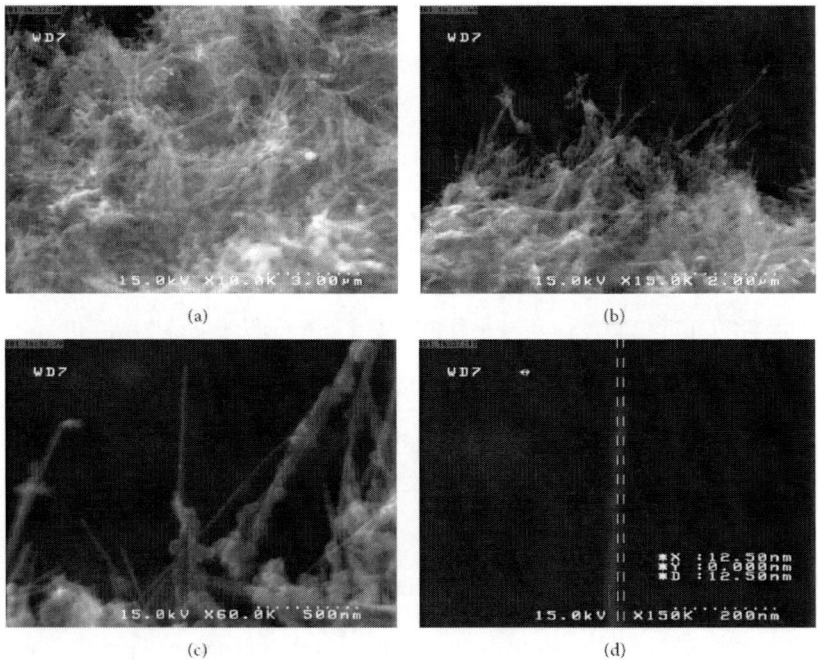

Figure 1.8. SEM images of the CNTs produced in the absence of a catalyst, (a) and (b) low magnification of the CNTs produced, while c) and d) illustrate high magnification of the CNTs [24].

The synthesis of the carbon nanotubes was done through the catalytic chemical vapour deposition (CVD) by utilizing ferrocene and molybdenum hexacarbonyl as catalysts [25]. There was a production of multi-walled carbon nanotubes with a diameter of 16-55 nm and a length of 1.2 μm in the presence of the iron-molybdenum alloy after methane decomposition reaction at 750°C for a duration of 35 min. Elsewhere in the literature [26], single-walled carbon nanotubes were produced by the arc discharge method. The carbon nanotube samples were fabricated at a constant helium pressure in the range of 500 to 700 Torr. The experiment was further done by applying the magnetic field in order to confine the discharge plasma. The fabricated samples, i.e., SWCNTs, were collected after a duration of a 180s-run of the arch discharge through different conditions in the presence and absence of the magnetic field. According to the TEM, images consisted of isolated SWCNTs and bundles. Aboul Enein et al. [27] produced MWCNTs by the catalytic pyrolysis of the LDPE waste. The LDPE pyrolysis and growth of the CNTs were done utilizing two separated reactors in both vertical and horizontal positions. The vertical reactor consisted of LDPE waste and H-ZSM-5 zeolite, while the Co/MgO catalyst charged into the horizontal reactor. The catalytic process of LDPE waste was done at various temperatures within the range of 350-600°C. The product of this process was non-condensable and condensable gases in the vertical reactor. The produced non-condensable gases were deposited into the vertical reactor and decomposed to the CNT in the presence of Co/MgO at the temperature of 700°C. To analyse the morphology of the produced carbon material, TEM was utilized. The MWCNTs produced by the catalytic process in the temperature range of 350-600°C showed dense MWCNTs with uniform diameters at all the investigated temperature(s). A combination of thin and thick MWCNTs with diameters less than 45 nm were revealed in all temperature range.

1.2.2. Graphite

Graphite is a natural material appearing in crystalline, lumpy, and amorphous forms and it is the most stable form of carbon on Earth [28]. It consists of layers of sp^2 hybridized carbon atoms (Figure 1.9) [29].

The layers are named the graphene sheets or graphene layers and are connected by weak Van der Waals forces, which provide graphite with its softness and flexibility. The presence of delocalized π-electrons plays an important role in terms of making graphite an excellent thermal and electrical conductor in the direction of the planes. The adjoining graphene sheets in the graphite are distanced from one another by 0.335 nm. Graphite is applied as a reinforcing material in various polymer matrices; however, graphite in its bulk state occurs in a layered form. As a result, it has to be separated and incorporated into a polymeric matrix in a dispersed manner. The separation of natural graphite results in various types of graphite. Three treatment methods are utilized in order for graphite to be modified. This modified graphite is called (i) graphene oxide (graphite oxide), (ii) graphite intercalated compounds (GICs), and (iii) expanded graphite. Graphene oxide is fabricated by treating the graphite flakes with oxidizing agents in order to introduce the polar groups on the surface of the graphite, thereby broadening the interlayer spacing of the graphene planes [30]. Figure 1.10 illustrates the model of graphene oxide.

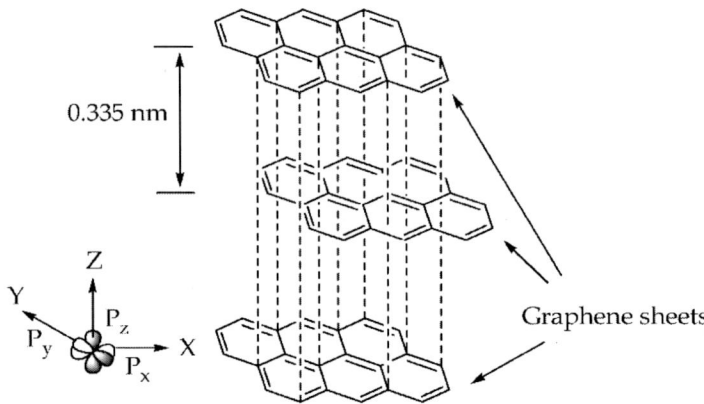

Figure 1.9. SP^2 hybridized carbon atoms of graphite [29].

The graphite intercalation compounds (GIC) are produced by the incorporation of the atomic or molecular layers with various chemical species between the layers of the graphite [31]. In the graphite intercalation compounds (GIC), graphene layers either accept or probably donate electrons to the intercalated species. GIC are key in the production of graphite, and potentially useful in various industries such as superconductors, anode materials, and catalysts [32]. The current usage of GIC is in the fabrication of the graphite nanoplatelets (GNP), expanded graphite, and single-layer graphene [32]. The fabrication of graphite nanoplatelets involves the heating of a graphite and potassium (K) to form

GIC. The GIC reacts with ethanol to cause exfoliation of the graphite layers. Expanded graphite is formed when the GIC is heated at a certain temperature or subjected to a microwave radiation. Normally, expanded graphite (EG) consists of nanosheets of various sizes ranging from 100 to 400 nm.

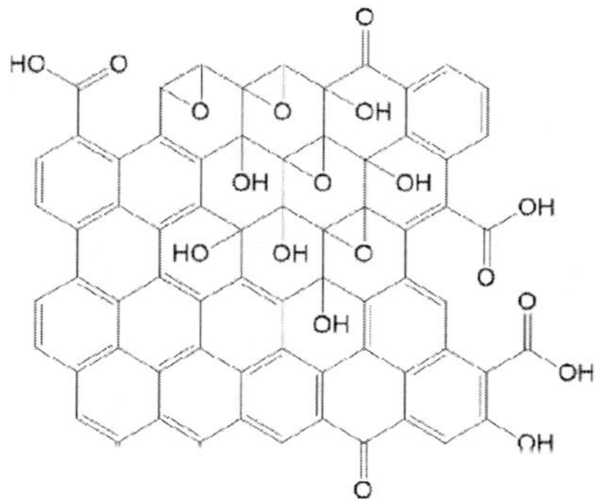

Figure 1.10. The Lerf-Klinowski model for graphene oxide [31].

1.2.2.1. Graphene: Synthesis

Graphene was first reported in 2004 and it is a single layer consisting of two-dimensional carbon atoms in a hexagonal lattice arrangement. Graphene has been utilized in various applications, including energy storage batteries, super capacitors, fuel cells, and solar cells. Graphene is one of the allotropes of the carbon. The other well-known allotropes of carbon include nanotubes, fullerenes, and graphene oxide. Several methods have been reported for the synthesis of graphene. It is noted that some of the methods are identical to those that are utilized for the synthesis of carbon nanotubes. Methods such a chemical vapour deposition (CVD) [33], chemical synthesis and mechanical cleaving are some of the methods used for the synthesis of graphene (see Figure 1.11). According to Figure 1.11, the synthesis methods are divided into the top-down and bottom-up synthesis methods.

The synthesis methods are differentiated into top-down and bottom-up methods. The top-down and bottom-up synthesis methods are characterized by the number of layers, thickness and nature as well as the average size of graphene materials [33]. Micromechanical exfoliation is known as the Scotch tape or a typical peel-off method. The fabrication of graphene is done with the assistance of the adhesive tape to catalyse or force the graphene layers to be separated [34-35]. During this process, multiple layers of graphene are attached to the tape after the peeling process, and with further peeling the graphene flakes split into more layers. The simplest method for the production of graphene has been recognized as the electrochemical techniques of graphite exfoliation [33]. Different forms of graphite such as graphite foils, rods, graphite powders and rods are electrodes used in both aqueous and non-aqueous electrolytes in order to bring about the expansion of the electrodes. The electrodes are named Cathodic (negative) and Anodic (positive) electrodes. It is reported elsewhere in the literature [36] that pure graphite electrodes, in combination with the polysodium-4-styrenesulfonate (PSS) solubilized in deionized water, have been used to incarnate the electrolyte. A couple of minutes after the electrolysis process, there was a production of the black material at the anode. During the fabrication process, the experiment was carried out for 4 hours in order to isolate the product from the cell. Pyrolysis is another method that has been utilized for the production of the graphene layers [37]. The simplest synthesis of graphene by pyrolysis is the thermal decomposition of silicon carbide (SiC).

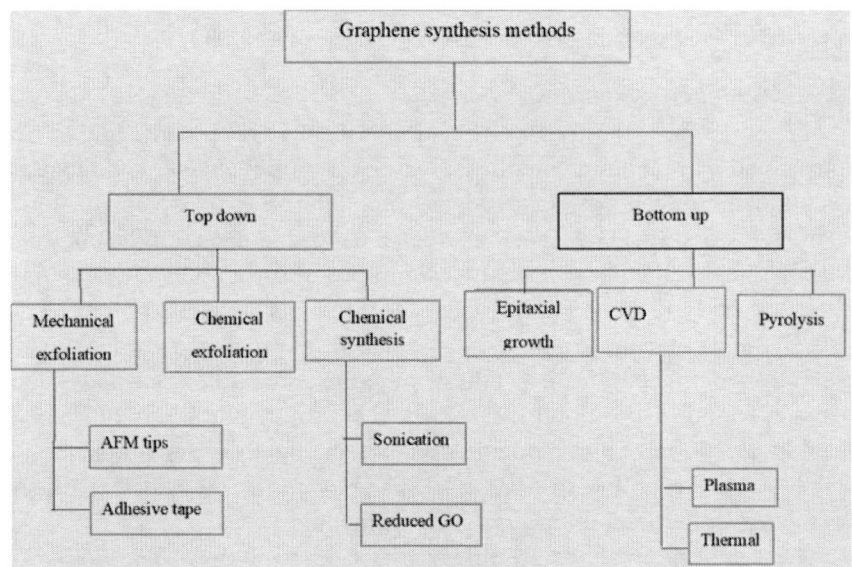

Figure 1.11. An illustration of various methods for graphene synthesis.

During this process, it is expected that at higher temperatures, Si is desorbed, as a result leaving the C atom, which produces a couple of graphene layers. The common method for the production of the graphene layers is chemical vapor deposition (CVD). The method utilizes various conditions such as the temperature, pressure, and gas flow under a chamber. During the CVD process for production graphene, various substrates are needed for the growth of graphene such as nickel (Ni), iron (Fe), copper (Cu) and stainless steel. Furthermore, two carbon sources are utilized, i.e., methane and acetylene. The purpose of gases such as hydrogen and argon in the CVD method is to remove any unwanted materials on the metal catalyst. There are various types of CVD methods, as illustrated in Figure 1.12. Various CVD methods are currently used, and they are categorized into seven types, which include pressure, temperature, gas flow state, wall/substrate, activation/power source, depositing time, and nature of precursor [38].

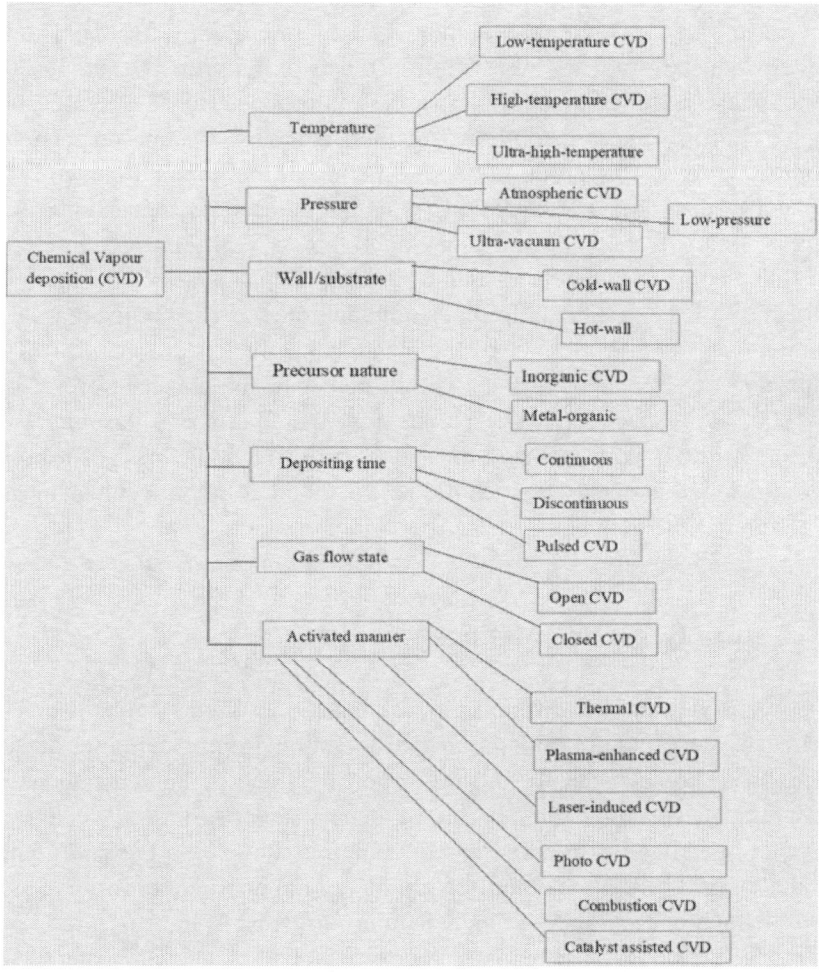

Figure 1.12. Various classifications of CVD.

Dhaouadi et al. [39] report on the synthesis of the monolayer graphene by chemical vapour deposition utilizing methane as a carbon source. The fabrication and synthesis of graphene occurred on a centimetre-sized copper substrate in the presence of air at a temperature of 180°C. The substrate, i.e., copper, was then heated by the electromagnetic induction in a constant atmosphere consisting of argon and hydrogen up until a steady temperature of approximately 1050°C. The reported results showed that an oxidized surface could possibly decrease the graphene nucleation density, as a result enhancing the graphene domain sizes. The produced graphene was analysed by Raman spectroscopy and scanning electron microscopy (SEM). The size of the produced graphene was within a range of 1 μm to 10 μm. In 2008, liquid phase exfoliation (LPE) was first introduced for the synthesis of graphene [40]. This is a top-down method whereby a stable dispersion of graphene layer can be produced by exfoliation of the graphite. Several steps are involved in the LPE method, which includes the dispersion of the graphite in a solvent, followed by exfoliation and finally purification. Gu et al. [41] report on the ultrasound-assisted, liquid-phase exfoliation for fabrication of the graphene. In this study, an aqueous deoxycholate (NaDC) solution was employed for exfoliation and the dispersion stability of the graphene. The frequency and net output power of the crusher were 20 kHz and 100 W, respectively. The results revealed that the ultrasonic waves played a critical role in terms of the size as well as the thickness distribution. The ultrasound-assisted deoxycholic acid sodium solution resulted in good exfoliation of the graphene.

1.3. Future Recommendations

It is realised that majority of the catalysts, modifiers and solvents are chemicals, which might be a problem from an environmental point of view. There is an urgent need for the utilization of green methods using natural products. There is the production of nanoparticles using a green route; however, the chemical route seems to be dominating at the present moment. By the green synthesis method for producing nanoparticles, we refer to the use of bioactive agents, which include micro-organisms, biowastes, and plant materials. Green synthesis in the current global crisis with reference to environmental impact seems to be a favoured route due to its advantages when compared with chemical and physical methods. The advantages of green synthesis include environmental friendliness and non-toxic methods of producing nanoparticles. However, there are minor disadvantages associated with the green synthesis method, such as longer reaction time [42].

1.4. Conclusion

In order to have carbon-based materials with specific properties, synthesis method(s) may be used whereby there is an altering of parameters such as catalyst, reaction temperature and reaction time, etc. Various synthesisation methods are used for the fabrication of carbon-based fillers, which include laser ablation, chemical vapor deposition, electrolysis, and arch discharge. Depending on the method used, various morphologies were obtained from various methods of synthesisation. The difference in morphologies was found to be affected by factors such as the content of the catalyst, growth time, and growth temperature. During the synthesis of the carbon nanotubes, it became evident that the presence of the catalysts is important, which in most cases is utilized for their synthesis. For graphene, it became evident that amongst all the synthesis methods used for its synthesis, chemical vapour deposition (CVD) was the preferred method of synthesis.

References

[1] Patel KD, Singh RK, Kim, H-W. Carbon-based nanomaterials as an emerging platform for theranostics. *Materials Horizons* 2019; 6:434-469.

[2] Kumar N, Kumbhat S. "Carbon-based nanomaterials". In book: *Essentials in Nanoscience and Nanotechnology,* Book authors Narendra Kumar and Sunita Kumbhat, New Jersey: John Wiley and Sons; 2019. p.189-236.

[3] Liu M-S, Lin M C-C, Huang, IT and Wang, C-C. Enhancement of thermal conductivity with carbon nanotube for nanofluids. *International Communications in Heat and Mass Transfer 2005*; 32(9):1202-1210.

[4] Papageorgiou DG, Kinloch IA, Young RJ. Mechanical properties of graphene and graphene-based nanocomposites. *Progress in Materials Science* 2017; 90:75-127.

[5] Japić D, Kulovec S, Kalin M, Slapnik J, Nardin B, Huskić M. Effect of expanded graphite on mechanical and tribological properties of polyamide 6/glass fibre composites. *Advances in Polymer Technology* 2022; Volume 2022: Article ID 9974889.

[6] Liu C, Wu W, Drummer D, Wang Y, Chen Q, Liu X, Schneider K. 2021. Significantly enhanced thermal conductivity of polymer composites via establishing double-percolated expanded graphite/multi-layer graphene hybrid filler network. *European Polymer Journal* 2021; 160:110768.

[7] Jena KK, AlFantazi A, Mayyas AT. Efficient and cost-effective hybrid composite materials based on thermoplastic polymer and recycled graphite. *Chemical Engineering Journal* 2022; 430:132667.

[8] Goyal M, Goyal N, Kaur H, Gera A, Minocha K, Jindal P. Fabrication and characterisation of low-density polyethylene (LDPE)/multi walled carbon nanotubes (MWCNTs) nano-composites. *Perspectives in Science* 2016; 8:403-405.

[9] Liu X, Kuang W, Guo B. Preparation of the rubber/graphene oxide composites with in-situ interfacial design. *Polymer* 2015; 56:553-562.

[10] Zhang Q, Wang J, Zhang B-Y, Guo B-H, Yu J, Guo, Z-X. Improved electrical conductivity of polymer/carbon black composites by simultaneous dispersion and interaction-induced network assembly. *Composites Science and Technology* 2019; 179:106-114.

[11] Danilov PA, Lonin AA, Kudryashov SI, Makarov SV, Mel'nik NN, Rudenko AA, Yurovskikh, VI, Zayarny DV, Lednev VN, Obraztsova ED, Pershin, SM, Bunkin, A.F. Femtosecond laser ablation of single-wall carbon nanotube-based material. *Laser Physics Letters* 2014; 11:106101.

[12] Bell MS, Teo KBK, Lacerda RG, Milne WI, Hash DB, Meyyappan M. Carbon nanotubes by plasma-enhanced chemical vapor deposition. *Pure and Applied Chemistry,* 2006;78:1117-1125.

[13] Subrahmanyam KS, Panchakarla LS, Govindaraj A, Rao CNR. Simple method of preparing graphene flakes by an arc-discharge method. *The Journal of Physical Chemistry C Letters* 2009 113:4257-4259.

[14] Gangoli VS, Godwin MA, Reddy G, Bradley RK, Barron AR. The state of HiPco single walled carbon nanotubes in 2019. *C Journal of Carbon Research* 2019; 5:65 (1-9).

[15] Sinclair RC, Suter JL, Coveney PV. Micromechanical exfoliation of graphene on the atomistic scale. *Physical Chemistry Chemical Physics* 2019; 21:5716-5722.

[16] Li L, Zhou M., Jin L, Liu L, Mo Y, Li X, Mo Z., Liu Z, You S, Zhu H. 2019. Research progress of the liquid-phase exfoliation and stable dispersion mechanism and method of graphene. *Frontiers in Materials* 2019; 6:325.

[17] Bleu Y, Bourquard F, Gartiser V, Loir A-S, Caja-Munoz B, Avila J, Barnier V, Garrelie F, Donnet C. Graphene synthesis on SiO_2 using pulsed laser deposition with bilayer predominance. *Materials Chemistry and Physics* 2019; 238:121905.

[18] Sharms HR, Ghanbari D, Salavati-Niasari MS, Jamshidi P. 2013. Solvothermal synthesis of carbon nanostructure and its influence on thermal stability of polystyrene. *Composites Part B: Engineering* 2013; 55:362-367.

[19] Aqel A, El-Nour K.M.M.A, Ammar RAA, Al-Warthan A. Carbon nanotubes, science and technology part (I) structure, synthesis and characterisation. *Arabian Journal of Chemistry* 2012; 5:1-23.

[20] Maheswaran R, Shanmugavel BP. A critical review of the role of carbon nanotubes in the progress of next-generation electronic applications. *Journal of Electronic Materials* 2022; 51:2786-2800.

[21] Eatemadi A, Daraee H, Karimkhanloo H, Kouhi M, Zarghami N, Akbarzadeh A, Abasi M, Hanifehpour Y, Joo SW. Carbon nanotubes: properties, synthesis,

purification, medical applications. *Nanoscale Research Letters* 2014; 9:393 (1-13).

[22] Cao TT, Ngo TTT, Nguyen VC, Than XT, Nguyen BT, Phan NM. Single-walled carbon nanotubes synthesized by chemical vapor deposition of C_2H_2 over an Al_2O_3 supported mixture of Fe, Mo, Co catalysts. *Advances in Natural Sciences: Nanoscience and Nanotechnology* 2011; 2: 035007 (1-5).

[23] Chrzanowska J, Hoffman J, Malolepszy A, Mazurkiewicz M, Kowalewski TA, Szymanski Z, Stobinski L. Synthesis of carbon nanotubes by the laser ablation method: effect of laser wavelength. *Physica Status Solidi B* 2015; 252:1860-1867.

[24] Mohammad MI, Moosa AA, Potgieter JH, Ishmael MK. Carbon nanotubes synthesis via arc discharge with a yttria catalyst. *International Scholarly Research Notices* 2023; Volume 2013: Article ID 785160 (1-7).

[25] Yahyazadeh A, Khoshandam B. Carbon nanotube synthesis via the catalytic chemical vapor deposition of methane in the presence of iron, molybdenum, and iron-molybdenum alloy thin layer catalysts. *Results in Physics* 2017; 7:3826-3837.

[26] Keidar M, Levchenko I, Arbel T, Alexander M, Waas AM, Ostrikov K.K. Magnetic-field-enhanced synthesis of single-walled carbon nanotubes in arc discharge. *Journal of Applied Physics* 2008; 103: Article ID 094318 (1-7).

[27] Aboul-Enein AA, Arafa EI, Abdel-Azim SM, Awadallah AE. Synthesis of multiwalled carbon nanotubes from polyethylene waste to enhance the rheological behavior of lubricating grease. *Fullerenes, Nanotubes and Carbon Nanostructures* 2021; 29(1): 46-57.

[28] Kasálková NS, Slepička P, Švorčík V. Carbon nanostructures, nanolayers, and their composites. *Nanomaterials* 2021;11: 2368.

[29] Hizam SMM, Al-Dhahebi AM, Saheed MSM. Recent advances in graphene-based nanocomposites for ammonia detection. *Polymers* 2022: 14(23):5125.

[30] Sengupta R, Bhattacharya M, Bandyopadhyay S, Bhowmick AK. A review on the mechanical and electrical properties of graphite and modified reinforced polymer composites. *Progress in Polymer Science* 2011; 36:638-670.

[31] Jiříčková A, Jankovský O, Sofer Z, Sedmidubský D. Synthesis and applications of graphene oxide. *Materials* 2022;15(3):920.

[32] Salvatore M, Carotenuto G, De Nicola S, Camerlingo C, Ambrogi V, Carfagna C. Synthesis and characterization of highly intercalated graphite bisulfate. *Nanoscale Research Letters* 2017; 12: 167.

[33] Mbayachi VB, Ndayiragije E, Sammani T, Taj S, Mbuta ER, Khan AU. Graphene synthesis, characterization and its applications: *A review. Results in Chemistry* 2021; 3:100163.

[34] Novoselov KS, Jiang D, Schedin F, Booth TJ, Khotkevich VV, Morozov SV, Geim, AK. Two-dimensional atomic crystals. *PNAS* 2005; 102:10451-10453.

[35] Edwards RS, Coleman KS. 2012. Graphene synthesis: relationships to applications. *Nanoscale* 2012; 5:38-51.

[36] Luheng W, Tianhuai D, Peng W. 2009. Influence of carbon black concentration on piezoresistivity for carbon-black-filled silicone rubber composite. *Carbon* 2009; 47(14):3151-3157.

[37] Hibino H, Kageshima H, Nagase M. Graphene growth on silicon carbide. *NTT Technical Review* 2010; 8(8):1-5.
[38] Saeed M, Alshammari Y, Majeed SA, Al-Nasrallah E. 2020. Chemical vapour deposition of graphene-synthesis, characterisation, and applications: A review. *Molecules* 2020; 25:3856.
[39] Dhaouadi E, Alimi W, Konstantakopoulou M, Hinkov I, Abderrabba M, Farhat S. 2023. Graphene synthesis by electromagnetic induction heating of oxygen-rich copper foils. *Diamond and Related Materials* 2023; 132:109659.
[40] Hernandez Y, Nicolosi V, Lotya M, Blighe FM, Sun Z, De S, McGovern IT, Holland B, Byrne M, Gun'Ko YK, Boland JJ, Niraj P, Duesberg G, Krishnamurthy S, Goodhue R, Hutchison J, Scardaci V, Ferrari AC, Coleman JN. High-yield production of graphene by liquid-phase exfoliation of graphite. *Nature Nanotechnology* 2008; 3:563-568.
[41] Gu X, Zhao Y, Sun K, Viera CLZ, Jia Z, Cui C, Wang Z, Walsh A, Huang S. Method of ultrasound-assisted liquid-phase exfoliation to prepare graphene. *Ultrasonics Sonochemistry* 2019; 58:104630.
[42] Ying S, Guan Z, Ofoegbu PC, Clubb P, Rico C, He F, Hong J. 2022. Green synthesis of nanoparticles: current developments and limitations. *Environmental Technology and Innovation* 2022; 26:102336.

Chapter 2

PLA Matrix: Synthesis Route, Structure, and Properties

Abstract

The structure, properties and modification of the PLA play a key role in the applications of the biopolymer. For example, low or medium molecular weight is utilized in applications such as medicinal applications for drug release and implants. However, a high molecular PLA is mostly used in a disposable consumer goods. In order to alter with the molecular weight of PLA, either to high or low molecular weight PLA, end groups of PLA play a critical role in this regard. Various components such as carboxylic acid (COOH) and amino groups are incorporated into the PLA matrix to alter the properties of the biopolymer. Furthermore, various synthesis methods for the production of PLA are also discussed. Various factors that affect the synthesis of PLA, such as the type of catalyst, reaction time and precursors, are also highlighted in depth.

Keywords: synthesis, lactide, biopolymer(s), polymerization, isomeric forms

2.1. Introduction

More research efforts have been dedicated to the utilization of biodegradable polymeric materials in different applications. Biopolymers are termed as the organic substances that are found in natural sources. The name 'biopolymer' comes from the Greek words *bio* and *polymer*, which symbolize nature and living organisms [1]. The biopolymers are known to be biocompatible and biodegradable, which enhance their widespread applications such as in emulsions, edible films, food packaging, drug transport materials and medical implants. The majority of the macromolecules are biopolymers comprising nuclei acids, proteins, carbohydrates, and lipids [2]. Biopolymers play an important role in nature [3-4]. They are key in terms of performing roles/functions such as storage of energy preservation and cellular construction. There are three categories that are obtained from renewable resources:

- Polymers produced from agro-resources (i.e., starch and cellulose)
- Polymers fabricated from microbial production, viz. polyhydroxy-alkanoates
- Chemically synthesized, whereby the monomers are produced from the agro-resources (i.e., PLA)

Table 2.1. Selective biopolymers with their sources [5]

Polymer	Source/Method	Example of bacteria: used for synthesis
Cellulose-based plastics	Biopolymer of glucose	30% of the biomass fabricated after extraction of algal oil known to consist of cellulose
Poly-lactic acid (PLA)	Polymerization of lactic acid	Bacterial fermentation
Bio-polyethylene	Ethylene fabricated from ethanol	Bacterial fermentation
Poly esters	Biomass	Bacteria such as *E. coli*

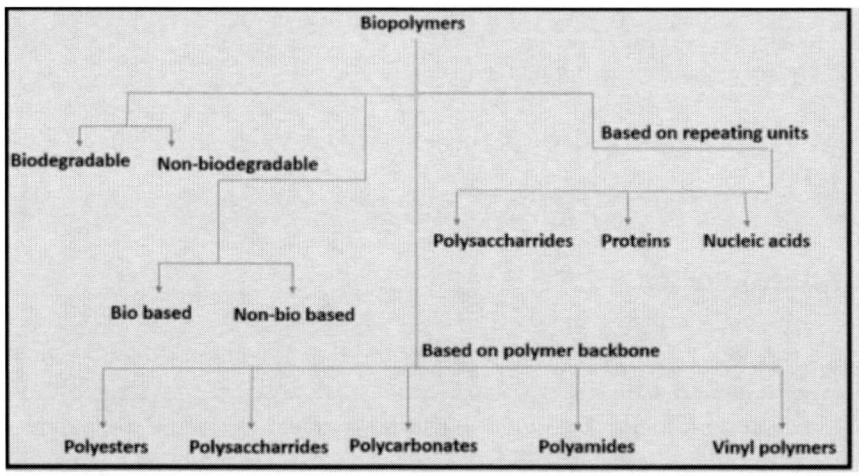

Figure 2.1. Classification of biopolymers based on their origin.

Biopolymers, which are fabricated from micro-organisms, need specific nutrients as well as specifically controlled environmental conditions [3]. In most cases their production occurs directly via fermentation or by chemical polymerization of the monomers. Biopolymers produced from the bacteria occur through the defence mechanism and/or as a storage material [4]. The advantage of biopolymers is that they can be degraded by natural processes, enzymes, and micro-organisms; as a result it may be reabsorbed into the environment. In this regard, there is a promotion of the development by shifting our focus to the biopolymers, as there would be a reduction in carbon dioxide emission. Table 2.1 illustrates selective biopolymers and their resources, while Figure 2.1 shows biopolymers together with their origin.

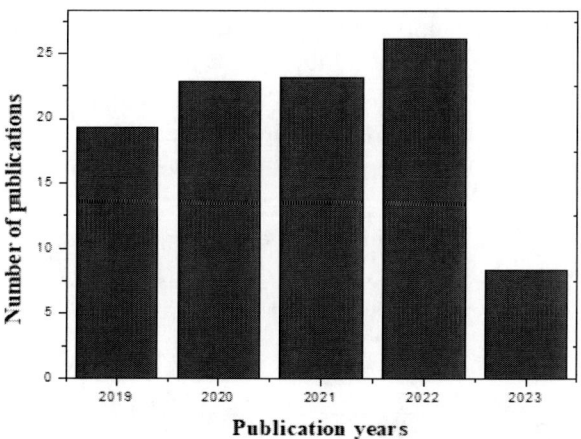

Figure 2.2. Top leading research areas by typing the search words "polylactic acid synthesis". 06 APRIL 2023.

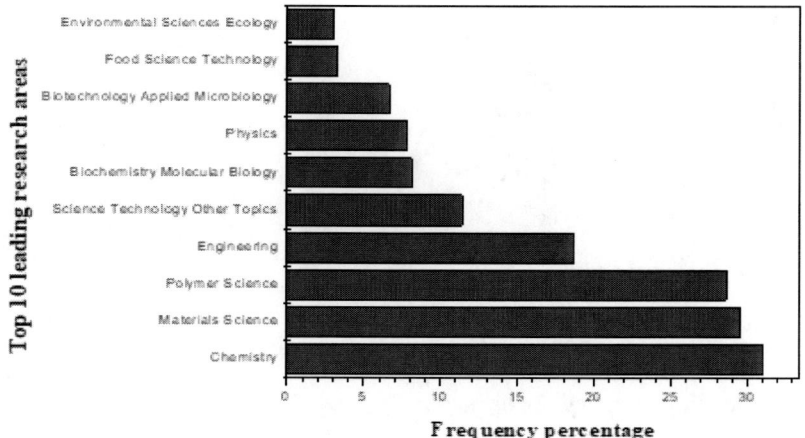

Figure 2.3. Top leading research areas by typing the search words "polylactic acid synthesis". 06 APRIL 2023.

One of the upcoming polymers with significant applications is polylactic acid (PLA). This is due to biocompatibility and biodegradability [6, 7]. PLA is produced from lactic acid (LA) and it can be transformed back into the LA monomer if it is hydrolytically degraded. The synthesis of PLA needs an employment of the catalysts with rigorous conditions such as temperature, pressure, and pH. The monomer in the form of the LA may be produced by chemical synthesis or microbial fermentation methods. The chemical route synthesis of LA has a major disadvantage, in the sense that there may be a production of the racemic mixtures. The fermentative method of LA production is the preferred method due to the production of optically pure $_L$ or $_D$ LA. This chapter discusses the synthesis and properties of PLA. Based on the "bibliometric analysis" by typing the search words "polylactic acid synthesis", there is a steady increase in the number of publications in the last five years (Figure 2.2).

Furthermore, it is apparent that the top leading research area in the synthesis of PLA is chemistry, followed by materials science and polymer sciences, while the 10th one on the list is environmental sciences ecology (Figure 2.3).

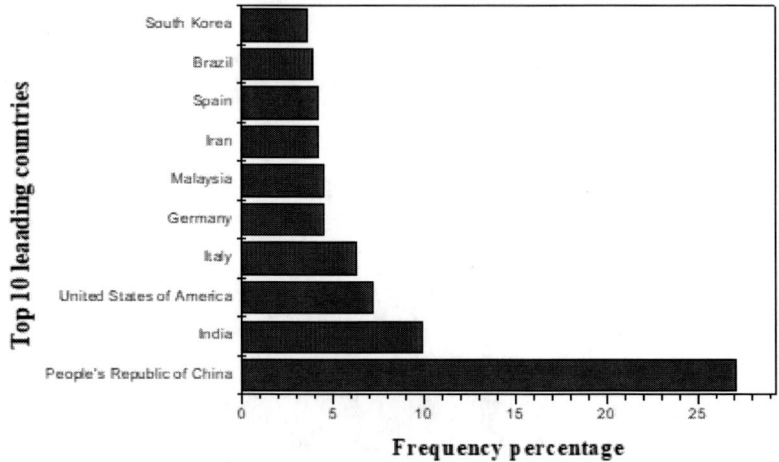

Figure 2.4. Leading country in terms of "polylactic acid synthesis". 06 APRIL 2023.

The leading country in terms of "polylactic acid synthesis" is China, followed by India and the United States of America (Figure 2.4).

Table 2.2. Manufacturer, price, and the type of lactide [8]

Manufacturer	L-Lactide			D-Lactide			D,L-Lactide		
	Product Number	Packaging (g)	Price (USD)	Product Number	Packaging (g)	Price (USD)	Product Number	Packaging (g)	Price (USD)
AK Scientific	V2322	500	907.00	2682BB	25	269	V3862	1	14.00
Alfa Aesar	L09031	25	66.40				L09126	10	34.70
Alichem				13076170	25	156.00			
Ambeed				A178884	25	102.00			
American Custom Chemicals Corporation	CCH003810	25	1212.75	CCH007427	0.005	502.59			
Apollo scientific	OR933335	25	115.00	OR959522	25	370.00			
Arctom							AS027325	25	60.00
Chem-Impex	35186	25	50.40					25	42.00
Crysdot	CD4500C285	25	91.00	CD11297333	100	371.00	CD45000014	500	510.00
Matrix Scientific		25	89.00				172406	25	89.00
Medical Isotopes, Inc.	65805	250	2200.00						
Sigma-Aldrich	367044	25	73.00				303143	25	71.70
SynQuest Laboratories	2H25-1-2T	25	184.00	2H25-1-29	25	592.00	2H25-1-2U	25	96.00
TCI Chemical	L0115	25	55.00				L0091	25	62.00
TRC	L113600	25	150.00	L113605	1	1135.00	L113518	0.5	70.00

Table 2.3. Selective physical properties of the lactide [8]

Physical Properties	L-Lactide	D-Lactide	Meso-Lactide	Rac-Lactide
Molecular weight (g/mol)	144.12	144.12	144.12	
Optical rotation in degrees	−260	+260		
Specific rotation (polarimetry toluene, 25 °C)	(−287)–(−300)°	(+287)–(+300)		(−1)–(+1)°
Appearance (visual test)	White crystals			
Melting point (°C)	95–100	95–100	53–54	122–128
Boiling point (°C)	255			142
Heat of fusion (J/g)	146		118; 128	120–170 (DSC 10 °C/min)
Heat of vaporization (kJ/mol)	63			
Solid density (g/mL)	1.32–1.38		1.32–1.38	
Liquid viscosity (mPas) (110 °C)	2.71			
Liquid viscosity (mPas) (120 °C)	2.23			
Liquid viscosity (mPas) (130 °C)	1.88			

2.2. Methods of Obtaining Lactide

Lactide is known as the monomer produced from lactic acid dehydration, followed by a subsequent prepolymer depolymerization process. Lactide is best defined as a cyclic ester, which is an intermediate product in the fabrication of the PLA. This monomer is known as an important monomer of the production of high-molar mass PLA [8]. Due to the importance of this particular monomer, i.e., lactide, there is a high price increase in lactide due to its demand commercially (Table 2.2).

Purified lactide depends on various parameters such as the reaction conditions and the oligomer composition. Factors such as time, temperature, pressure, molecular weight of the oligomer, concentration and type of the catalyst have to be optimized in order to obtain the high lactide yield. The lactide possesses two stereo centres, namely L-and D-lactide. The preferred synthesis of lactide is the L-lactide, since the L-lactide is bioabsorbable and biocompatible. The synthesis of the lactide is produced from the condensation of the lactic acid. The formation of the lactide takes place in two steps, namely the prepolymer derived from the lactic acid dehydration. The lactide is obtained from the depolymerization process. The formation of lactide from both steps occurs at temperatures above 200°C, which is an indication of higher energy utilization. The reverse engineering is another route that is utilized for the production of the lactide. In this type of production, the depolymerization of the oligomeric PLA takes place in reactions called back-biting and end-biting. The back-biting method/reaction produces a cyclic compound through the intramolecular reactions between the hydroxylic and the ester group of the oligomer chain.

2.3. The Properties of the Lactide

Lactide has scientifically been proven to have three isomeric forms, i.e., (S, S)-L, (R, R)-D, and (S,R)/(R,S)-meso, with the L-lactide being the preferred one commercially [8-11]. The optical activity of the lactic acid plays a critical role in terms of controlling the isomerism of the fabricated lactide, which controls the properties of the final product or polymer. It was noted generally that the physicochemical properties of the lactide does not differ among the isomeric forms. It was further realized that the polymerization reactions result in different properties for various isomerisms of the lactide in the production of PLA; characteristics such as the melting point, mechanical properties, degradation rate, extent of crystallization, and crystallization rate are all affected. Furthermore, it has been recognized that the melting temperature is an important property of the lactide due to the ROP occurring at temperatures above the lactide melting temperatures [8, 12]. The lactide melting temperatures range from the 95-98°C (D-lactide) and L-lactide, 53-54°C for meso-lactide, while the *rac*-lactide has a melting temperature within the range of 122-126°C. The ROP method of production allows the control over the distribution of comonomers in the polymer chain. This method further produces low polydispersity, high molar mass and high-end groups in comparison to the utilization of the polycondensation. The enantiomerically pure monomers (L- and D-lactide) results in the formation of polymer chain isotactic structure, whereas the *rac*-lactide utilized as monomer may be used for obtaining isotactic structure [13]. Based on the thermodynamic properties, lactide is regarded as a six-membered ring that is easily polymerizable [8]. Table 2.3 summarizes selective properties of the lactide.

2.4. Synthesis of PLA

Polylactic acid (PLA) has been synthesized by various polymerization methods, including (i) polycondensation of lactic acid monomer, (ii) ring-opening polymerization of Lactide and (iii) azeotropic dehydration condensation reaction, and (iv) Enzymatic polymerization and chain extension from the lactic acid [8]. Figure 2.5 illustrates different methods for synthesis of PLA. Amongst the above-mentioned synthesis methods, direct polymerization and ring-opening polymerization are the preferred methods for production of PLA [14].

Viamonte-Aristizábal et al. [13] report on the synthesis of polylactic acid based on the reactive extrusion ring-opening polymerization of L-lactide in the presence of 1,12-dodecanediao (DDD) or di(trimethylol propane) as initiators. In this study, the authors explored the chemical side of the reactive extrusion (REX) process, which is utilized for the synthesis and modification of PLA-based materials. The aim of the study was to use 1,12 dodecanodiol (DDD) and di-trimetilolpropane (DTMP), together with octoate and triphenylphosphine (TPP) for the fabrication of high molecular weight L-PLA through a reactive extrusion method. Polymerization of PLA by the reactive extrusion method was done by a co-rotating, twin-screw extrude. The dimension of the extruder was reported to have a barrel dimension of 28.3 mm and a length-to-diameter (L/D) ratio of 56. Due to the utilization of the extrusion process for the synthesis of PLA, processing conditions used were 165-210°C, a speed of 30 rpm and a throughput of 4kg/h. The reactive extrusion for synthesis PLA with 1-12 dodecanodiol (PLA_DDD) has responsive values of 31-32%, a pressure of 28-29 bar and a melting point of 172°C. However, response values for PLA fabricated from di(trimethylol propane) (PLA_DTMP) were recorded as 29-31% torque, a pressure of 32-33 bar with a melting point of 173°C. The appearance of the PLA_DTMP was seen as slightly yellowish when compared with the PLA_DDD counterpart. The structural analysis of the two synthesized was done by the Fourier transform infrared (FTIR) method. The PLA_DDD and PLA_DTMP revealed the characteristic absorption peaks around 2997-2944 cm^{-1}, which corresponds with the asymmetric as well as the symmetric vibrations of Csp^3 -H for H-CH$_2$. The H-C vibrations were also observed around 2880 cm^{-1}, while the C—O vibration of cyclic lactones was observed at 1745 cm^{-1}. The bands at 1445, 1357, and 1383 cm^{-1} were ascribed to the asymmetric and symmetric vibration of C-H from the CH$_3$. Furthermore, the conversion reported in this study for the synthesis of the two PLA was revealed as 98%. Pivsa-Art et al. [15] report on the synthesis of poly (D-lactic acid) by utilizing a two-step direct polycondensation process. The polymerization process was done in combination with esterification utilizing *p*-Toluenesulfonic acid in the absence of a metal catalyst. The two-step polycondensation involves mely-polymerization as well as solid-state polymerization. In this process there is a formation pre-polymer of D-lactic acid having the degree of polymerization within the range 4-7. Secondly, the pre-polymer was further modified into a fine powder and underwent solid-state polymerization (SSP) under 10 torr with a temperature below the melting temperature (T$_m$). The synthesized PLA was analysed through various techniques such as the DSC, TGA, and molecular weight analysis. The produced poly(D-lactic acid) revealed a melting temperature of 177°C, with an average molecular weight of 33 000 Da accompanied by a decomposition temperature of 255°C. Horváth et al. [16] report on the ring-opening polymerization of the PLA. The plasticization of the poly(L-lactic) acid was also investigated by incorporating glycerol-dioleate. In the process of polymerization, various temperatures and process times were used in order to reach the optimum production of the product. The produced PLA sample was analysed by FTIR, DSC, and gel permeation chromatography (GPC). Figure 2.6 shows the synthesis route of PLA.

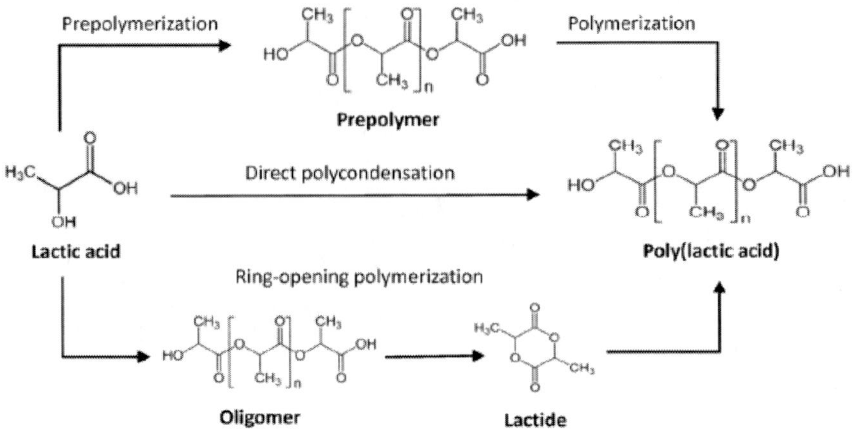

Figure 2.5. Various synthesis route for PLA [14].

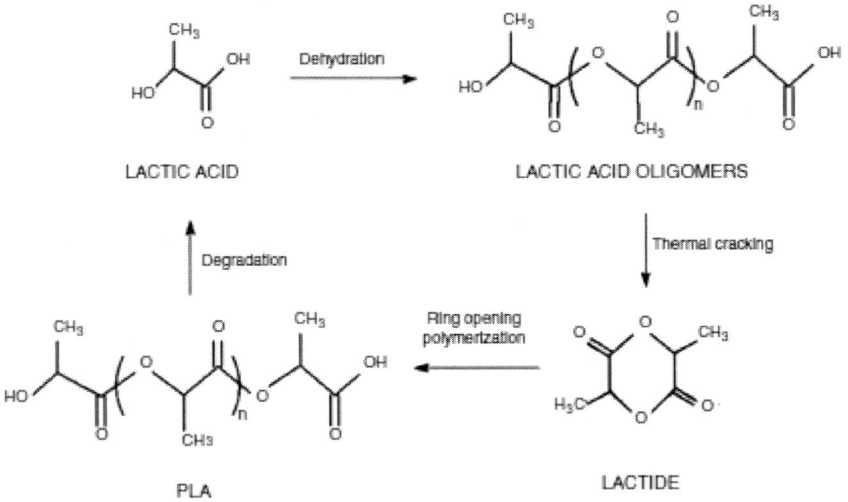

Figure 2.6. Synthesis of route of PLA [16].

In order to plasticize the produced PLA polymer, 6 g of the polymer was mixed with 2 mass/% (0.12 g) of the glycerol-dioleate and the mixture was heated up to 160°C for 1 hour. Both polymers were reported to have the same functional groups (Figure 2.7) as polylactic acid; as a result it can be concluded that polylactic acid was produced.

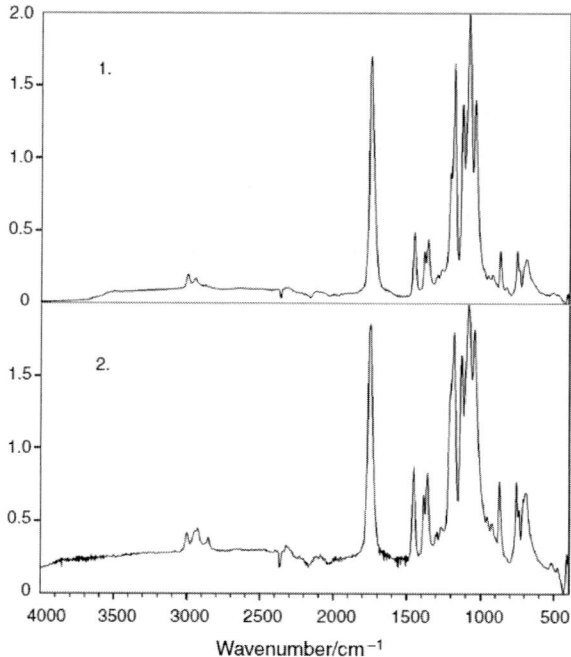

Figure 2.7. FTIR spectra of: 1: PLLA fabricated by ROP method and 2: Plasticized PLLA synthesized by polycondensation [16].

The glass transition temperature (T_g) of the PLLA produced by the ROP method was recorded to be 55-63°C, while the melting temperature (T_m) was 175.5°C. The T_g of the plasticized PLLA was found to be 41.9°C. The reduction in T_g is expected, since the plasticizer is able to enhance the motion of the segmental polymer chains at lower temperatures. Pholharn et al. [17] report on the ring-opening polymerization of the PLA by using a macro-initiator. The well-known initiators for PLLA are 1-dodecanol and various alcohols during ring-opening polymerization. However, in the study by Pholharn et al. [17], macro-initiator in the form of polybutylene succinate (PBS) and stannous octoate was used as catalyst for fabrication of PLA. The reason for the utilization of PBS as a macro-initiator is because PBS was found to be miscible with the lactide. As a result there is an improvement in the efficiency of the ring-opening polymerization. The produced PLLA was characterized by various techniques such as the differential scanning calorimetry (DSC), and proton nuclear magnetic resonance (^{1}H-NMR). The PLLA synthesized from bulk L-lactide in nitrogen at temperatures such as

120, 160 and 200°C for various hours, i.e., 4, 8, 12 and 16 h. There was an observation of the double melting peak in all of the synthesized PLLA. The first melting was ascribed to the melting peak (within the range of 70-98°C) of the lactide, while the peak in the range between 117-214°C was due to the melting of the PBS. The main melting peak for PLLA was found at a temperature of 150-170°C. It was deduced from the DSC results that the optimum condition for 0.01 mol% PBS was 160°C for 4 hours producing the highest PLLA crystallinity (i.e., 80%) and melting point (i.e., 165°C). The ^{1}H-NMR spectrum of the PLLA (initiated at PBS 0.001mol%, 160°C and 4 h) revealed signals at 1.56, 5.16 and 4.36 ppm, which normally correspond with methyl proton (-OCH$_3$), and methane as well as the terminal methane proton (HOCH), respectively within the L-lactide (Figure 2.8).

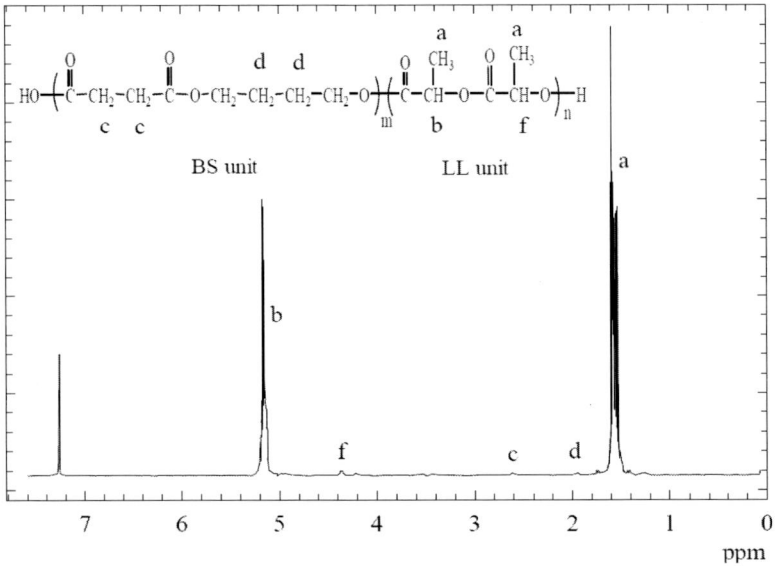

Figure 2.8. The 1H-NMR spectrum of PLLA with the use of PBS 0.001mol% as an initiator [17].

Char et al. [18] utilized the magnesium and zinc hetero-bimetallic complexes together with diol ligands for the synthesis of PLA. In their work, active as well as stereoselective initiators were used for polymerization of highly heterotactic PLA. The two catalytic complexes consisting of magnesium and zinc complexes 2 and 3, along with a primary alcohol were applied for ring-opening polymerization of the lactide. Both zinc and magnesium catalytical systems resulted in high performances for polymerization of the lactide, while there was an ease of the control in relation to the molecular weight, tacticity, and molecular weight distribution.

2.5. Incorporation of Other Groups into PLA Matrix: Efforts to Functionalize PLA

There are important factors that affect the properties of PLA such as the chemical components and compositions. PLA has various limitations such as a poor toughness, being hydrophobic, few reactive side chain groups, and slow degradation rate [19]. PLA modification consists of hydrophilic monomers such as polyethylene glycol (PEG) and monomers consisting of functional groups such as amino and carboxylic groups [20]. Baimark et al. synthesized a bioplastic from poly(L-lactide)-b-polyethylene glycol-b-poly(L-lactide) by ring-opening polymerization [21]. The synthesis reaction for such a system is shown in Figure 2.9. The copolymer was synthesized at 165°C in a nitrogen atmosphere for a duration of 6 hours, with $Sn(Oct)_2$ used as a catalyst. Furthermore, there was an incorporation of the Joncryl® CE (chain extender) within the system with contents of 1.0, 2.0 and 4.0 phr. The samples were tested against various techniques such as the gel content, melt-flow index, intrinsic viscosity, thermal properties, and thermal decomposition. It was observed that the molecular size of the PLLA was directly associated with the gel-content values. The increase in gel content was observed with an increase with CE content. The CE with a content of 4.0 phr revealed the highest gel content, approximately 70.4%. This behaviour may be attributed to the cross-linking reaction that takes place between PLLA-PEG-PLLA and CE chains. Furthermore, there was a drop in the MFI with the addition of 1.0 to 2.0 phr in the system (Table 2.4). The observation was ascribed to the branching structures of the PLLA-PEG-PLLA.

The DSC technique revealed that the cold crystallization (T_{cc}) and glass transition (T_g) of PLLA (Table 2.5) were observed to shift to higher temperatures with an increase in CE content. This is due to the branching and crosslinking. One may realize that the presence CE played a negative role in terms of the chain mobility during the process from glassy to rubbery state as well as the cold crystallization.

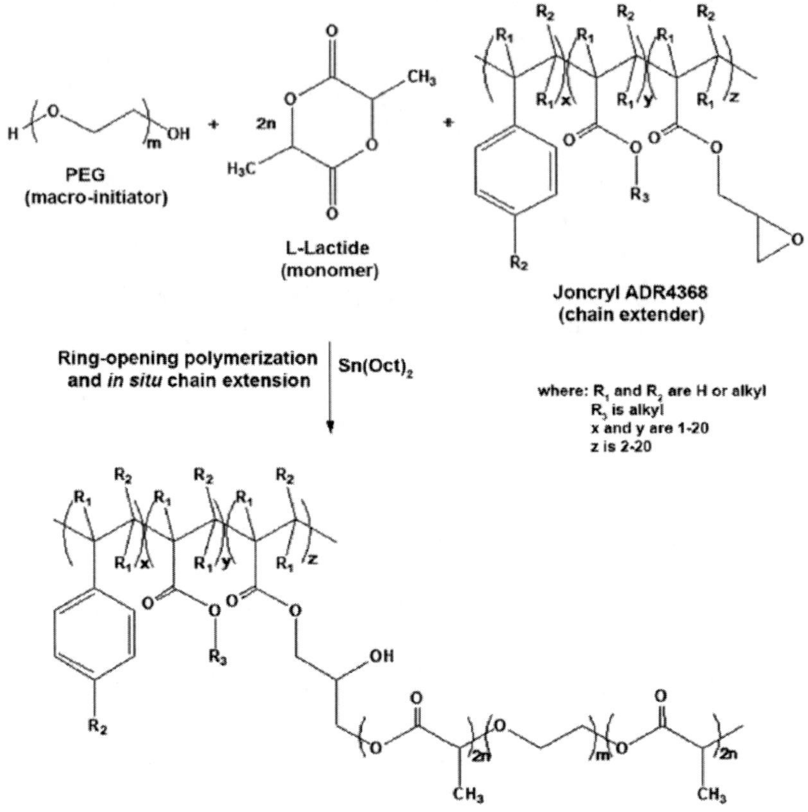

Figure 2.9. Synthesis route for PLLA-PEG-PLLA (ring-opening polymerization) [21].

Table 2.4. Gel content and MFI results of the PLLA-PEG-PLLA [21]

CE content (phr)	Gel content (%)	$[\eta]^a$ (dL/g)	Branching degree[a]	MFI (g/10 min)[b]
—	0.2 ± 0.1	1.2	—	—[c]
1	2.5 ± 1.1	1.7	1.42	51 ± 8
2	4.4 ± 1.5	1.9	1.58	26 ± 6
4	70.4 ± 6.8	2	1.67	12 ± 3

[a] Measured from soluble fraction of PLLA-PEG-PLLA in chloroform.
[b] Determined at 190°C under 2.16 kg load. [c] Could not measure.

Table 2.5. Summary of the DSC results from the heating scans [21]

CE content (phr)	T_g (°C)	T_{cc} (°C)	ΔH_{xx} (ϑ/γ)	T_m (°C)	ΔH_μ (ϑ/γ)	DSC-X_c (%)
—	29	—	—	170	44.3	57
1	30	70	14.1	164	37	29.4
2	31	75	14.1	161	30.8	21.4
4	32	76	13.8	160	26.8	16.7

Polylactic acid functionalized with carboxylic acid and aldehyde was reported by Icart and co-workers [22]. During the functionalization process of the PLA, salicylic aldehyde (SAl) and salicylic acid (SAc) were employed as co-initiators of the ring-opening polymerization. The catalyst used in this process was Tin (II) 2-ethylhexanoate. It was observed that the co-initiator and the catalyst molar ratio (viz 12/1) had a huge impact on the PLA functionalization. Aldehyde and carboxylic acid in PLA were reported to increase with an increase in co-initiator/catalyst molar ratio. The optimum conditions for functionalizing PLA with aldehyde and carboxylic acid were found at the co-initiator/catalyst molar ratio of 12/1. However, the co-initiators/catalyst molar ratios played no significant role on the molecular weight, while an increase in lactide/Sn ratio increased the molecular weight of the polymer. Generally, the content of the end groups (*viz* aldehyde and carboxylic) in the PLA depended on the ratios mentioned above. Reungdech et al. studied the functionalization of the polylactide with pol(ethyleneimine) by the process of *in situ* reactive extrusion [23]. The fabrication of the PLA-PEI was done by a two-step process: a PLA-maleate (PLA-MA) precursor with various compositions of the maleic anhydride, i.e., 1, 2, 3 and 4% by weight per weight (w/w). Secondly, the two types of PEI with different molecular weights, i.e., PEI_{800} (1-4%) and PEI_{25k} (1-2%), were incorporated into the PLA-MA system by ring-opening reaction. The prepared samples (PLA-MA and PLA-PEI) were analysed by Fourier transform infrared spectroscopy and nuclear magnetic resonance. In short, this study briefly discussed or investigated the functionalization of PLA into the PLA-PEI in order to introduce the amino groups in the PLA structure for application in antibacterial activity. According to [23], PLA is inert; as a result, the possible incorporation of the MA into the PLA before its synthesis to form PLA-PEI is through a reactive extrusion. The structure of the three samples

(PLA, PLA-MA, and PLA-PEI) were analysed by the FTIR. PLA revealed an absorbance peak at 3200-2800 cm^{-1} for the C-H stretching, C=O stretching around the 1746 cm^{-1}, 1180 cm^{-1} for C-O band, and 1451 cm^{-1} for C-H bending. All the functional groups observed for PLA were reported to be present for PLA-MA, except for an additional peak at 695 cm^{-1}, which was attributed to the anhydride. The grafting of PLA with MA was reported to affect the molecular weight (MW), with a decrease in MW with MA grafting. The reduction in MW is attributed to PLA undergoing a β-scission during grafting, with possible transition into smaller by products. Kitahara et al. [24] investigated the inclusion of carboxylic acid on the surface of a porous poly(lactic acid). The porous poly(lactic acid) was prepared by utilizing the so-called salt leaching and subsequent freeze drying. The carboxylic acid group (COOH) is inserted into the PLA matrix to ensure that the polar organic acid had any impact on the apatite attachment when compared with unmodified PLA. In order to modify the PLA with carboxylic acid, PLA was added to the sodium hydroxide at room temperature (20°C) for a period of 1 hour for hydrolyzation of the of the PLA scaffold (Figure 2.10).

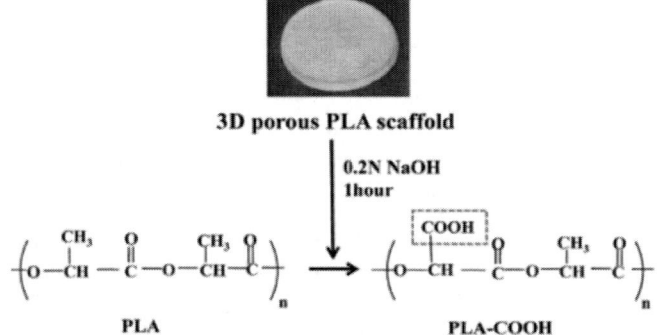

Figure 2.10. Schematic representation of the modified PLA by COOH [24].

The sample was further modified by rinsing it with double-distilled water together with hydrochloric acid and further again with double-distilled water. According to the cross-section of the SEM view, the pore size was revealed to be 15-30 μm inside the scaffold. After their incorporation into the Hank's balanced salt solution (HBSS) with a pH=7.4, for 3, 7, 14 and 28 days, there was obvious apatite precipitate in the PLA-COOH scaffold. The amount of precipitate depended hugely on the immersion time, as the prolonged time enhanced the precipitates (see Figure 2.11).

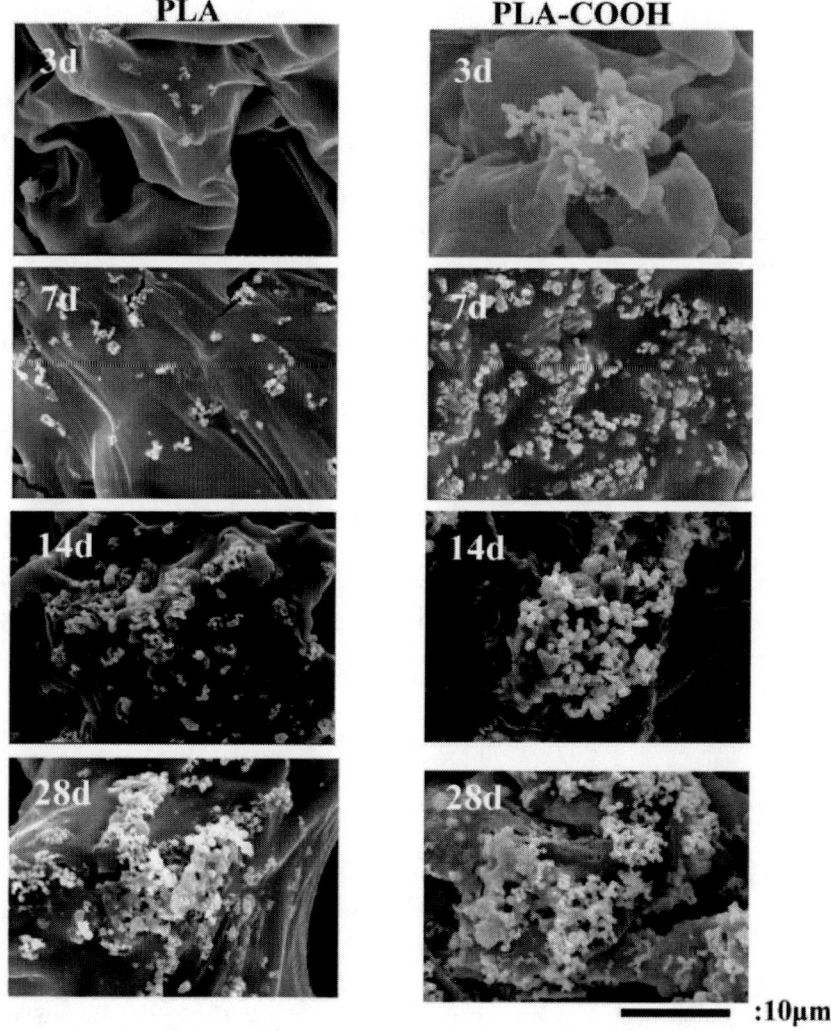

Figure 2.11. SEM images of the porous PLA and PLA-COOH scaffolds at various immersion days (i.e., 3, 7, 14 and 28) [24].

To further determine or investigate the rate and degree of degradation, water absorption of the PLA and COOH-PLA was also reported. Water adsorption of the sample was found to depend on the immersion time. The modified sample was found to increase in water until an optimum number of days, namely 28 days. Furthermore, there was a minimum of differences in terms of water adsorption between the PLA and COOH-PLA samples, which showed that the COOH had no effect on the rate and degree of degradation.

2.6. Future Recommendations

There is a need for the production of PLA by trying to explore inexpensive substrates and the use of high performance of microorganisms to improve the production efficiency of the lactic acid. To further reduce the production cost, it is suggested that the PLA be blended with the copolymers or other polymers to form blends. Furthermore, polymer blending is an effective and cost-effective way of improving the properties of PLA, more specifically the toughness. For decades, blending has been found to be an economic method of producing new materials with better properties than the neat polymer, which in turn widens the applications of the polymers. As result, blending is expected to enhance the performance and provide balanced properties of PLA with its blends, which in turn is expected to result in a low-cost and flexible product. Another factor that most researchers have to take into consideration while they are improving the toughness of PLA, is to employ or utilize a compatibilizer that is environmentally friendly, since most of the polymer blends are immiscible.

2.7. Conclusion

The chapter reported on the synthesis, modification, and properties of the PLA. The properties of the lactide were further discussed, since it is the main component for the synthesis of the PLA. PLA synthesis consists of multiple steps with the production of lactic acid and ends with polymerization. It was noticed that the preferred synthesis methods include (i) direct condensation polymerization, (ii) ring-opening polymerization, and (iii) azeotropic dehydration condensation. The final product and its molecular weight depend on factors such as the solvent, reaction temperature, polymerization process, catalyst, and the quantity of the impurities. It was also realized that the production of PLA is very expensive due to the high cost of the lactide monomer.

References

[1] Baranwal J, Barse B, Fais A, Delogu GL, Kumar A. Biopolymer: A sustainable material for food and medical applications. *Polymers* 2022; 14:983.

[2] Das A, Ringu T, Ghosh S, Pramanik N. A comprehensive review on recent advances in preparation, physicochemical characterization, and bioengineering applications of biopolymers. *Polymer Bulletin* 2023; 80:7247-7312.

[3] Huang S, Xue Y, Yu B, Wang L, Zhou C, Ma Y. 2021. A review of the recent developments in the bioproduction of polylactic acid and its precursors optically pure lactic acids. *Molecules* 2021; 26:6446 (1-16).

[4] Sukan A, Roy I, Keshavarz T. Dual production of biopolymers from bacteria. *Carbohydrate Polymers* 2015; 126: 47–51.

[5] Mohan S, Oluwatobi SO, Nandakumar K, Sabu T, Songca SP. "Biopolymers – Application in Nanoscience and Nanotechnology". In: *Recent Advances in Biopolymers, edited* by Farzana Khan Perveen, Rijeka, Croatia: IntechOpen; 2016. p. 47-72.

[6] Montané X, Montornes JM, Nogalska A, Olkiewicz M, Giamberini M, Garcia-Valls R, Badia-Fabregat M, Jubany I, Tylkowski B. Synthesis and synthetic mechanism of polylactic acid. *Physical Sciences Reviews* 2020; 5:20190102.

[7] Carvalho JRG, Conde G, Antonioli ML, Dias PP, Vasconcelos RO, Taboga SR, Canola PA, Chinelatto M.A, Pereira GT, Ferraz G. *Polymer Journal* 2020;52: 629-643.

[8] Cunha BLC, Bahú JO, Xavier LF, Crivellin S, de Souza SDA, Lodi L, Jardini AL, Filho RM, Schiavon MIRB, Concha VOC, Severino P, Souto EB. Lactide: Production routes, properties, and applications. *Bioengineering* 2022; 9:164.

[9] Masutani K, Kimura Y. 2015. "PLA Synthesis. From the Monomer to the Polymer". In *Poly(Lactic Acid) Science and Technology: Processing, Properties, Additives and Applications,* edited by Jiménez A, Pelzer MA, Ruseckaite RA, London, UK. The Royal Society of Chemistry; 2015.p. 1-36.

[10] Kolstad JJ. Crystallization Kinetics of Poly(L-Lactide-Co-Meso-Lactide). *Journal of Applied Polymer Science* 1996; 62:1079–1091.

[11] Trivedi AK, Gupta MK, Singh H. PLA based biocomposites for sustainable products: A review. *Advanced Industrial and Engineering Polymer Research 2023.* https://doi.org/10.1016/j.aiepr.2023.02.002 (In press).

[12] Groot W, Van Krieken J, Sliekersl O, de Vos S. 2011. "Production and Purification of Lactic Acid and Lactide". In *Poly (Lactic Acid): Synthesis, Structures, Properties, Processing, and Applications,* edited by Auras R, Lim L.-T, Selke SEM, Tsuji H, Hoboken, NJ, USA: John Wiley & Sons, Inc.; 2011. p. 3-16.

[13] Viamonte-Aristizábal S, García-Sancho A, Campos FMA, Martínez-Lao JA, Fernández, I. Synthesis of high molecular weight L-polylactic acid (PLA) by reactive extrusion at a pilot plant scale: Influence of 1,12-dodecanediol and di(trimethylol propane) as initiators. *European Polymer Journal* 2021; 161:110818.

[14] Hu Y, Daoud WA, Cheuk KKL, Lin CSK. Newly developed techniques on polycondensation, ring-opening polymerization and polymer modification: focus on polylactic acid. *Materials* 2016; 9:133 (1-14).

[15] Pivsa-Art S, Tong-Ngok T, Junnggam S, Wongpajan R, Pivsa-Art W. Synthesis of poly(D-lactic acid) using a 2-steps direct polycondensation process. *Energy Procedia* 2013; 34:604-609.

[16] Horváth T, Marossy K, Szabó T. Ring-opening polymerization and plasticization of poly(L-lactic) acid by adding of glycerol-dioleate. *Journal of Thermal Analysis and Calorimetry* 2022; 47:2221-2227.

[17] Pholharn D, Srithep Y, Morris J. Ring opening polymerization of poly(L-lactide) by macroinitiator. *AIP Conference Proceedings* 2019; 2065:030016.

[18] Char J, Brulé E, Gros PC, Rager M-N, Guérineau V, Thomas CM. 2015. Synthesis of heterotactic PLA from rac-lactide using hetero-bimetallic Mg/Zn-Li systems. *Journal of Organometallic Chemistry* 2015; 796:47-52.

[19] Rasal RM, Janorkar AV, Hirt DE. Poly(lactic acid) modifications. *Progress in Polymer Science* 2010; 35:338-356.

[20] Cheng Y, Deng S, Chen P, Ruan R. Polylactic acid (PLA) synthesis and modifications: a review. Frontiers of Chemistry in China 2009; 4(3):259-264.

[21] Baimark Y, Rungseesantivanon W, Prakymoramas N. Synthesis of flexible poly(L-lactide)-b-polyethylene glycol-b-poly(L-lactide) bioplastics by ring-opening polymerization in the presence of chain extender. *e-Polymers* 2020; 20:423-429.

[22] Icart LP, Fernandes E, Agüero L, Cuesta MZ, Silva DZ, Rodríguez-Fernández DE, Souza Jr FG, Lima LMTR, Dias ML. End functionalization by ring opening polymerization: influence of reaction conditions on the synthesis of end functionalized poly (lactic acid). *Journal of the Brazilian Chemical Society* 2017; 00:1-10.

[23] Reungdech W, Tachaboonyakiat W. Functionalization of polylactide with multibranched poly(ethyleneimine) by in situ reactive extrusion. *Polymer* 2022; 246:124746.

[24] Kitahara K, Fuse M, Fukumoto M. Effect of the Introduction of a Carboxylic Acid Group onto the Surface of a Three-dimensional Porous Poly (lactic acid) Scaffold on Apatite Deposition. *International Journal of Oral-Medical Sciences* 2011; 10(3):179-186.

Chapter 3

Preparation and Morphology of Carbon-Based Filler(s) Reinforced PLA

Abstract

Polylactic acid (PLA) is one of the polymers with a 'green' identity used in different sectors, such as packaging and the automotive, agricultural, textile and biomedical fields because of its distinctive properties (e.g., renewability, biodegradability and sustainability). PLA's limitations include poor mechanical and thermal properties, which restrict it from being employed in high-performance and long-term applications. These limitations can be addressed by utilizing a wide variety of nanofillers. Carbon-based fillers (CBF), which have exceptional properties, have drawn the most attention for their potential to improve the overall properties of PLA. The morphology and consequently the properties of PLA/CBF composites are influenced by the method of preparation. This chapter discusses various preparation techniques used to develop PLA/CBF composites.

Keywords: morphology, preparation, solution mixing, melt-spinning, melt-blending

3.1. Introduction

The rising concerns for our environment and strict governmental regulations pressurize the industries to develop eco-friendlier products [1–4]. Over the years, industrial and research communities have been working around the clock to develop advanced materials having a 'green' signature to avoid a negative impact on the environment. Biopolymers with their attractive attributes (e.g., biodegradability, sustainability and renewability) received tremendous interest as replacement for petroleum-based polymers [3, 5, 6]. Amongst these polymers, PLA received a great deal of interest because of its comparable properties to conventional polymers [1, 4, 7]. It is derived from natural sources, and thus it is biodegradable, bioabsorbable, and renewable, which afforded its use in packaging, pharmaceutical, textile, biomedical engineering and other industries. Additionally, its properties can be modified by changing crystallinity and molecular weight [1]. Its widespread use in long-term and high-performance applications is constrained by some inherited limitations. Poor mechanical performance and thermal properties are the main disadvantages that limit its materialization beyond biomedical and packaging applications. In this regard, the inclusion of different nanofillers into PLA is one of the suitable techniques to improve the overall properties of the resulting composite materials. A wide variety of fillers, including clay, nanocellulose, silica, graphene, carbon nanotubes, etc. have been incorporated into PLA to produce high-performance materials [2]. Carbon-based fillers have received the most attention among these fillers due to their distinctive properties, which include exceptional thermal stability, outstanding mechanical performance, high thermal and electrical conductivity, and many more [8].

The dispersion state, the amount of filler, and the interfacial adhesion between the components of the composite were all reported to be directly correlated with the properties of the resultant composite. These variables typically depend on the preparation technique (i.e., solution casting, in situ polymerization, melt-compounding), as well as whether PLA or a carbon-based filler has been modified. Additionally, the primary problem has been the concentration of the CBF within the PLA as a result of Van der Waals forces. In this chapter, the influence of production processes of PLA/carbon-based filler (CBF) on the morphology is discussed. The effect of modification of either PLA and/or CBF on the morphology of PLA/CBF composite is also covered.

3.2. Preparation of PLA/CBFs Composites

There are three commonly used methods to prepare PLA/CBFs composites, namely melt-blending, in situ polymerization and solution mixing [1]. Recent studies indicate that three-dimensional printing has been gaining attention as secondary melt-blending process to enable the application of PLA/CBFs composites in various fields. Nonetheless, melt-blending can be categorized into plain melt-mixing and single/two-screw extruder as discussed in the next subsections.

Table 3.1. Melt-blending of PLA composites

Formulation	Filler content	Preparation method	Functionalization	Highlights	Refs.
PLA/CNTs	0.5-3	• Melt-mixing • Melt-pressed	Acid treatment (-COOH)	Functionalization of CNTs promoted their dispersion	[9]
PLA/PU/CNTs	10	• Melt-mixing • Melt-pressed	Alkyl chain (C18) Alkyl chain + Ti(Obu)4	The modification of CNTs promoted their interaction with PLA via the formation of hydrogen bonding between the composite components	[10]
PLA/CNTs	05-5	• Melt-extrusion • Melt-spinning	• Ozonolysis for 60 minutes • UV irradiation for 15 minutes	Higher loadings led to breakage of the fibres due to strong filler-filler interaction	[11]
PLA/CNTs	0.5-3.0	• Melt-mixing • Compression moulding	• CNTs functionalized with –OH • Acrylic acid-g-PLA (PLA-OH)	CNTs were well dispersed at low contents but formed agglomerates at high content (3wt5). CNTS were well embedded in the matrix due to the formation of branched ester bonds resulting from CNT-OH and PLA-g-AA via condensation reaction. The strong interaction and better dispersion improved both mechanical and thermal properties of the composites	[12]
PLA/carbon black (CB)	1-7	Melt-compressing	-	Increase in content led to CB conduction network within PLA which improved electrical and mechanical properties	[13]
PLA/PBS/SCF/GO	SCF 3-9 GO 0.5	Melt-mixing Extrusion FDM-printing	-	The presence of GO improved interaction between PLA and SCF by filling the gaps between these components	[14]
CB/PLA	4-20	Melt-mixing Compression moulding	-	Optimal content of 8-12% CB was best to improved overall properties	[15]

3.2.1. Melt-Blending

Melt-blending is the commonly employed processing technique to fabricate PLA/CBFs composites, as reviewed in Table 3.1 [9, 10]. It is the most economically viable, eco-friendly and easily scalable preparation method. Melt-blending involves the addition of CBFs into molten PLA to allow the material to pass through the compounder one or more times to afford the desired product. The dispersion of the fillers is primarily influenced by processing conditions, such as mixing speed, time and temperature. It is important to note that the state of the filler dispersion and distribution are often poor when using melt-blending, compared to other processing techniques, i.e., solution mixing and in situ polymerization. In addition, the secondary processing techniques, including compression moulding, injection moulding, three-dimensional (3D) printing, etc., are often employed after pelletizing of the composite strands to fabricate the anticipated product (Table 3.1).

3.2.1.1. Melt-Mixer
Melt-mixing is usually used on a laboratory scale to afford the preparation of small sample portions for quick analysis [9, 10, 12]. A melt-mixer comprised a bowl for feeding the materials consisting of three independent heating zones and two counter-rotating blades. The benefits of using a melt-mixer are the ability to change the blades-type to enable thorough mixing, depending on the material, small portions of the pre-constituents required (i.e., 40-70 g), ease of control over the programmable temperature, time and shears. This techniques are employed to evaluate the possibility of using the as-prepared materials in large, industrial-scale production.

The modification of CNTs is usually adopted to promote the interaction and dispersion of the filler within the polymeric matrix. Lin et al. [9] prepared PLA/CNT composites using an internal mixer with co-rotating speed of 50 rpm at 180°C for 5 minutes. The authors purified CNTs and then functionalized it with a long alkyl chain via transesterification between –COOH and stearyl alcohol to improve their dispersion. The unmodified CNTs resulted in a lot of entangled clusters within PLA because of strong Van der Waals forces between CNTs. Purified alkyl functionalized CNTs and purified alkyl functionalized CNTs in the presence transesterification agent (Titanium butoxide $(Ti(OBu)_4)$-based composites were homogeneously distributed within the host matrix. The functionalization and presence of TA promote grafting of CNTs onto PLA chains, and thus strong interaction as presented by fewer cracks on the fractured surface when compared to unmodified CNTs. This was attributed to limited wetting of the polymer to unmodified CNTs, hence leading to brittle fracture. TEM images showed that CNTs were embedded within PLA for functionalized CNTs in the presence of TA, while alkyl-grafted and unmodified CNTs were situated on the surface of the polymer, which drastically affected the surface resistivity of the composites. It was noticed that alkyl-grafted CNTs performed better than other composites due to better distribution. The latter resulted in the formation of conductive path within the composites that afforded superior electrical conductivity.

The modification of CNTs using ozonolysis and UV irradiation to improve the interaction between host polymeric material and CNTs was reported in the literature [10]. It was found that the modification of CNTs introduced surface functionalities that were responsible for the formation of hydrogen bonds when blended with PLA/PU blends. Thus, the mechanical properties, storage modulus and glass transition temperature were improved for modified CNTs-based composites when compared to unmodified CNTs-based composites. In addition, thermal and electrical conductivity was significantly improved. Elsewhere, the modification of both PLA and CNTs was employed to improve the overall dispersion and interfacial adhesion between CNTs and PLA [12]. The authors used dicumyl peroxide as an initiator to graft acrylic acid onto PLA (PLA-g-AA), while CNTs were purified using an acid combination before being subjected to a reaction with thionyl chloride to introduce chlorocarbonyl groups (CNTs-COCl). Then, CNTs-COCl were reacted with 1.6 hexadiol to afford multi-hydroxyl functionalized CNTs (CNTs-OH). It was noticed that the wettability of CNTs-OH within PLA-g-AA was improved when compared to neat PLA and bare CNTs-based composites. CNTs were well-dispersed with agglomerates being formed at higher loadings. This was attributed to the dehydration of the –COOH from PLA-g-AA and the –OH groups from CNTs-OH, which led to the formation of ester linkages. In most cases, PLA or CNTs are modified to make up for the surface energy differential between them. Without adjustment, this energy disparity causes the carbon fillers to clump together. In addition, there is poor compatibility between the components of composites. The amount of filler was also reported to play a major role on the resulting morphology of composites prepared by a melt-mixer. Within this context, CB/PLA composites were prepared using melt-mixing, followed by compression moulding [15]. It was found that appropriate CB content is required to attain the desired properties. The increase in CB content up to 12% was sufficient to improve the overall properties, due to good dispersion and interfacial adhesion. Beyond 12%, the agglomerates prevail, which affect the performance of the composites negatively. In general, the presence of these fillers further improves the stiffness of PLA, which limits its applications. Therefore, the content of carbon filler has to be optimized, depending on the intended application. The rigidity of the filler contributes to the brittleness of the resulting PLA-based composites. Some of the reports indicate that tougher polymers and/or plasticizers are introduced to enhance toughness while facilitating processability of PLA [2, 4, 16, 17].

The brittle nature of PLA can be resolved by using plasticizers in the presence of the filler [16, 17]. Different plasticizers, including oligomeric lactic acid, tributyl citrate, diethyl methyl malonate, glycerol, citrate ester, and poly(propylene glycol) have been employed to improve the toughness of PLA [17]. However, the tensile strength and modulus of PLA are adversely impacted by the presence of these plasticizers. The use of nanofillers serves as an appropriate remedy to counteract such behaviour [16]. CB was incorporated into PLA/acetyl tributyl citrate (ATBC) or poly(1,3-butylene adipate) (PBA) blends to improve the mechanical properties of the resulting composite [16]. CB were agglomerated like the islands in the sea without any form of connection with one another, although there was interaction between PLA and CB. The presence of plasticizers allowed for improved dispersion and a strong interaction between PLA and CB. The processability of PLA was also enhanced, which resulted in decreasing the size of CB agglomerates.

3.2.1.2. Extrusion Method

The extruder primarily consists of the feeding zone, kneading zone and the heating zone [18]. The materials are introduced into the feeding zone, and then subjected to high shearing, temperatures and pressures are attained from the kneading and the heating zones to afford an as-prepared product. Different parameters, such as feeding rates, screw speed, temperature, screw length-to-diameter ratio (L/D) and die shape can be controlled to achieve the anticipated product [18].

Villmow et al. [18] investigated the effect of extrusion conditions on the dispersion of CNTs in PLA. The authors started by preparing a masterbatch composed of PLA and CNTs using a co-rotating twin-screw extruder with a barrel length of 900 mm (L/D = 36) and screw diameter of 25 mm. the resulting pelletized masterbatch was diluted with neat PLA and introduced into two different screw configurations. The first screw is made up of a melting zone o 5 left-handed conveying elements and 3 left-handed blocks of 5 kneading discs (+45°) and neutral staggering angles, i.e., 90°. The second screw had a melting zone assured by kneading elements, followed by a right-handed conveying element. In the case of masterbatch, two contents of CNT, i.e., 7.5% and 15% were chosen to evaluate the extent of dispersion using the same processing conditions. It was noticed that a higher dispersion index of 82.6% was attained for 15% loading when compared to 7.5% loading, with only a 63.4% dispersion index. The calculated number of agglomerates were 23.7 counts/mm^2 for 15% CNTs loading; meanwhile for 7.5% CNTs loading is 129.6 counts/mm^2. On the other hand, the cluster sizes were bigger for higher loadings when compared to lower loading, demonstrating that CNTs were more dispersed at higher loading with the remaining particle forming larger clusters. This phenomenon is due to an increase in viscosity with an increase in CNTs. At higher viscosities more shear stress is applied to the polymer melt, which in turn breaks down the agglomerates. Similar behaviour is observed when the screw speed is increased to maximum speed.

CB/PLA composites prepared using extrusion as a primary preparation method are processed further to afford the desired product [19, 20]. Similar to other carbon nanofillers, extrusion is usually adopted to prepare a master batch to improve the dispersion of the filler before the extrusion process. PLA is then added to the as-prepared mixture, and subsequently extruded and pelletized [19]. Cardoso et al. [19] prepared CB/PLA and hybrid (CB/alumina) using an extrusion process to afford nanocomposites that can be used for 3D-pronting applications. Prior to this, a master batch was prepared and then diluted with neat polymer for extrusion, followed by pelletization. The pellets were extruded using a benchtop single-screw extruder to afford filaments with 1.75 mm for 3D printing purposes. It was concluded that the presence of CB improves the dispersion of alumina, resulting in better thermal, mechanical and tribological properties. In the case of filler size, the smaller-sized filler with large surface area enhances the interaction between the filler and PLA. Elsewhere, the size of the CB was reported to play an essential role in their interaction with PLA [20]. It was noticed that poor interfacial adhesion was obtained when high surface-area CB was used as reinforcing filler for PLA. The utilized CB had a high surface area with a BET value of 1228 m^{2}g^{-1}.

3.2.1.3. Compression Moulding

Compression moulding is often used to manufacture large products (Table 3.1). It can also be used to prepare composites, which can further be processed by secondary melt-blending techniques. PLA and carbon black (CB)-based composites were prepared using compression moulding for the production of 3D-printing filaments by Masiuchok et al. [13]. PLA was ground to a powder before physically mixed CB particles. The obtained mixture was then introduced into a mould followed by melt-pressing at 200°C and 20 MPa pressure for 5 minutes to form a cylinder with a diameter of ~ 9 mm and ~12 mm in length (Figure 3.1). It was found that a conductive network between CB was obtained when the content of CB ranged between 2.5-7%, which improved both the electrical conductivity and mechanical properties of the resulting composite materials.

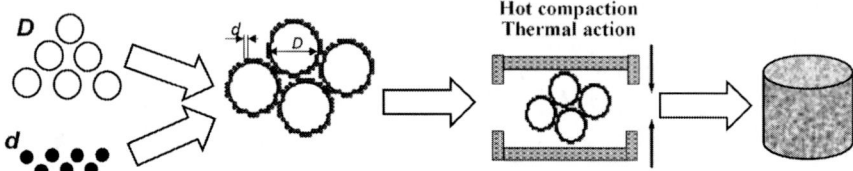

Figure 3.1. Schematic presentation of compression moulding. Reprinted with permission from Masiuchok, et al. [13]. Open Access.

It is worth mentioning that compression moulding can be classified as secondary preparation method for both melt-mixing and solution-mixing processes. In the case of melt-mixing, the pellets or lumps of composites are melted and pressed into the desired mould to afford the desired product for intended application [12] (Table 3.1). Solution casting often leads to discontinuous films with poor mechanical performance, and in order to ensure complete removal of solvent and continuous film formation, melt-pressing is often utilized. This will be discussed thoroughly in the next sections.

3.2.2. Secondary Melt-Blending

3.2.2.1. 3D Printing

Additive manufacturing is one of the most-used techniques to manufacture customized products [21-23]. It has the capability of producing complicated products from various materials without generating large amounts of waste when compared to traditional methods (e.g., compression moulding). This technique prints 3D structures through layer-by-layer deposition according to the CAD model. There are different printing techniques such as selective laser sintering (SLS), laminated (LOM), stereolithography (SLA), and Fused deposition modelling (FDM) [23]. Amongst these techniques, FDM is the most commonly used method due to its potential to print complex designs, mass customization and inexpensiveness. In this case, the 3D structure is printed from a filament that is feed into a heater which melts the polymeric material to afford its deposition in layer-layer fashion from printhead (Figure 3.2). Therefore, this technique use thermoplastic polymers and their composites as feeding stock (filaments) to afford the printing of different designs. PLA-based materials have received tremendous interest to produce filaments for 3D printing due to their biodegradability, biocompatibility and possible renewability. Carbon nanomaterials are often incorporated into PLA to overcome some of its limitations, such as low mechanical strength, and slow crystallization rate, which affect its crystallinity, and thus its properties [21].

Carbon nanotubes have been introduced into PLA to fabricate filaments that can result in better adhesion between the printed layers [21, 22]. The resulting composites are recognized by enhanced electrical conductivity and better mechanical performance [21]. It is worth mentioning that there is optimal CNT content in order to attain better overall performance of the as-prepared composites.

In most cases, the incorporation of neat CNTs results in their limited interaction and dispersibility within the PLA matrix. The surface modification of CNTs and/or the use of compatibilizer can be adopted to overcome those limitations. Bortoli et al. [21] investigated the inclusion of CNTs into PLA to fabricate fused filaments. They noticed that the treatment of CNTs with nitric acid introduced defects with additional oxygen functional groups responsible for strong interaction as well as good dispersion of the CNTs within the matrix. It is worth mentioning that the pre-preparation of the composites is often carried out before the filament can be produced. In this case, the authors prepared composites using an internal mixer by first introducing PLA pellets, followed by CNTs. The obtained mixture was ground and introduced into a single-screw extruder to obtain modified and unmodified composites. The functionalization of CNTs was found to improve the interfacial adhesion between the printed layers when compared to unmodified and neat PLA, which showed voids between the layers. It was noticed that the printed layers of PLA and PLA/CNTs samples were thinner when compared to that of PLA/functionalized CNTs due to the detachment of the layers during fracturing. This indicates that the modification of CNTs is required in order to improve adhesion between the layers and hence improve the printability of the PLA-based materials. The thermal properties were also similar for all the printed layers, indicating that all the samples experience the same thermal history.

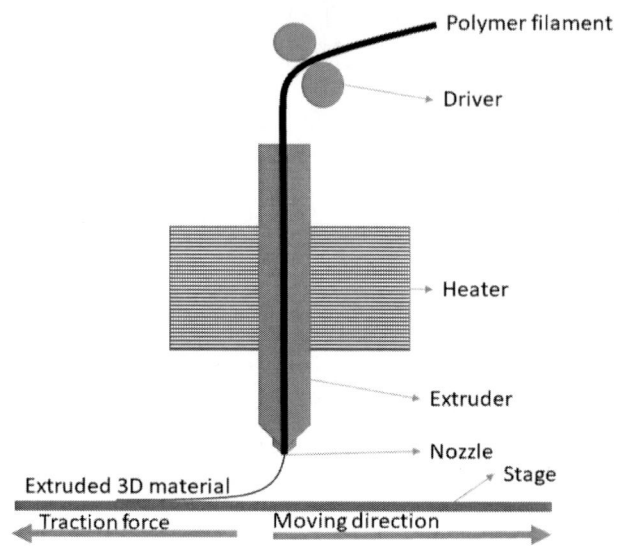

Figure 3.2. Schematic presentation of FDM printing method.

Similar to other carbon-based materials, graphene was also employed as reinforcing agent in PLA composites for fused deposition filament manufacturing [24-26]. The introduction of graphene-based materials into PLA increased mechanical performance and conductivity of the resultant printed products. Depending on the system under study, the maximum content of graphene material is required to achieve the preferred properties, depending on the intended application [25]. The inclusion of tougher polymers into PLA to overcome its brittleness could also counteract the rigid carbon filler, which further enhances the brittleness of PLA [25]. In most cases, the compatibilizers are introduced into the system to improve interfacial adhesion between the blends' components [25]. On the other hand, the presence of the conductive fillers improves the thermal conductivity of the resulting nanocomposite materials [26]. The higher the thermal conductivity led to faster cooling of the deposited layer which, in turn, limits the interaction between the layers [26]. Therefore, the content of the filler is important, since it influences the thermal conductivity of the resulting composite material.

PLA reinforced with carbon fibres has been one of the commonly used filaments in 3D printing [23]. These filaments are available in the market under different trading names, such as XT-CF20 (ColorFabb), CCTREE, Protopasta, etc. [27]. The carbon-fibre-based filaments are known for their unique features such as high modulus and ease of printing, and thus are often used when designs with high modulus and smooth surfaces are preferred. Carbon-fibre-reinforced PLA filaments are often made up of short fibres. However, a number of studies based on the use of continuous carbon-fibre as reinforcement of PLA are reported in the literature [27, 28]. For instance, in their work, Tian et al. [28] incorporated continuous carbon fibres (CCBFs) into PLA filament during printing to afford an integrated preparation of different designs through FDM. The authors investigated the influence of the various processing parameters (temperature of liquefier, layer thickness, feed rate of filament, hatch spacing, and transverse movement speed) on the temperature and pressure, since they play a crucial role in the printing process, thus affecting the mechanical performance of the 3D-printed structure. It was noticed that a temperature liquefier of 200-230°C provides good impregnation of CCBF into PLA, whereas bonding strength between the lines and layer can be achieved when the layer thickness is between 0.4 mm and 0.6 mm with a hatch spacing of ~0.6 mm. The resulting printed objects exhibit remarkable mechanical performance; hence they can be applied in aviation and aerospace.

Hybridization of short-carbon-fibre (SCF)-reinforced PLA filament with CCBF was studied by Maqsood and Rimašauskas [27]. It was noticed that CCBF had limited interaction with the PLA/SBF composite, which resulted in poor mechanical performance when compared to PLA/CCBF composites. The latter had better stress transfer between the components due to strong interfacial adhesion between PLA and the CCBFs. The hybrids showed some voids that could also be contributed to such poor mechanical performance. Elsewhere, hybridization of carbon fibres and graphene nanofillers was adopted to improve the overall properties of the resulting 3D-printed structures [14]. The authors physically mixed PLA, carbon fibres and graphene nanofillers in the presence of PBS as a toughening agent, followed by melt-blending. The resulting composite was crushed and extruded to afford preferable properties for FDM printing technology. The presence of

GO improved interaction between PLA and carbon fibres without showing any gaps, which was the case for SCF/PLA composites.

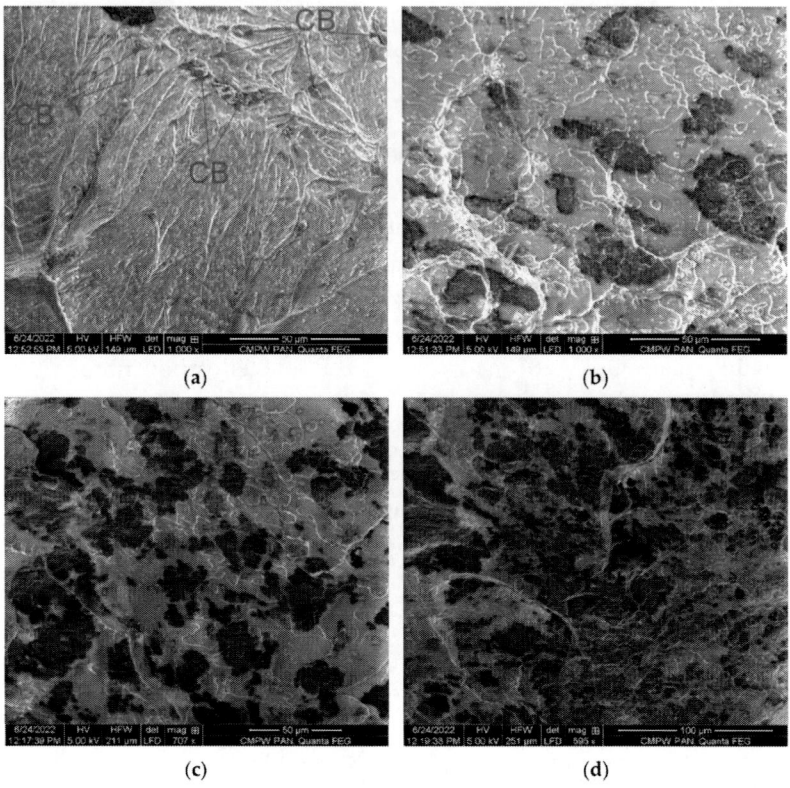

Figure 3.3. SEM images of transverse cross-section of 3D-printed CB/PLA composites: (a) 1%CB, (b) 2.5%, (c) 5% and (d) 7%CB. [13] Open access.

Similarly, carbon black-reinforced PLA filaments are commercially available [13, 29]. Like any other carbon-based filler, the conductive network of CB is crucial for the application of the resulting composite material [13, 29]. Recent studies by Masiuchok et al. [13] investigated the printability of CB/PLA to afford conductive 3D-printed structures. It was found that the conductive network of CB within PLA became more pronounced with CB content (Figure 3.3), which resulted in an increase in electrical conductivity of the printed 3D structures. The use of CB to promote the dispersion of other fillers has been reported in the literature [19]. In this case, strong CB interaction with the nanofillers serves as mode to promote their dispersion within the PLA matrix [19]. Such a composite displays better properties when compared to single-filler, reinforced composites due to synergistic effect.

3.2.2.2. Melt-Spinning

Melt-spinning is usually used as secondary melt-processing technique to afford melt-spun fibres with diameters ≥10 μm [2, 11, 30-32]. In this context, the PLA/CBF composites are produced using an extrusion method, followed by pelletization. Subsequently, the pellets are introduced into a melt-filament technique to produce nanocomposite fibres with their properties that are directly dependent on the spinning temperature, extrusion rate, take-up rate and draw ratio [18, 30-32]. Besides the fact that PLA is melt-spinnable even at high take-up velocity [30-32], the presence of the filler, content of the filler and interaction between the filler and PLA play a crucial role in the resulting properties of the melt-spun [11]. The preparation of PLA/CNTs using the melt-spinning method is often started with an extrusion method [11, 18]. The CNTs are often introduced into PLA by preparing a master batch followed by a dilution process to afford the inclusion of the desired CNTs content [11, 18].

Pötschke and co-workers from the Leibniz Institute of Polymer Research Dresden investigated the effect of CNTs loading (i.e., 0.5-5%) and take-up rates (20 to 100 m min^{-1}) on the resulting melt-spun fibres [11]. The authors claim that the spinnability of the composite was preserved up to 3% CNT loading, regardless of the take-up velocity employed. Beyond 3%, CNT concentration led to an increase in filler-to-filler interaction, which considerably decreased the melt's flowability and hence negatively affected the composite material's ability to spin. It was noticed that fibre breakage was observed for 5% CNT loading, even when the lowest take-up velocities (20 m min^{-1}) were used. The interaction of the fillers reduced the drawability of the fibres, which caused the diameters of the fibres to become uneven. Optical microscope images indicated that CNTs were well-dispersed within PLA, while TEM demonstrated that there were few agglomerates of loosely packed CNTs embedded within the matrix, and their maximum diameter was ~500 nm. In addition, the CNTs were highly oriented, with their clusters being stretched along the fibres drawing direction. This was confirmed by Raman measurements using the ratio of D- and G-band in parallel and perpendicular directions. Besides the fact that the D-band was slightly higher than the G-band, both were increased with an increase in take-up velocity, indicating that the CNTs were highly orientated.

3.2.3. Solution Mixing

Solution mixing is an often-used method to fabricate polymer-based composites, as summarized in Table 3.2 [33-35]. It is the simplest method without the use of special instruments, with a huge possibility of scale-up. The major concerns about solution mixing is the cost associated with the solvent and vapours, unless the solvent is water. Unfortunately, PLA can only dissolve in more polar solvents, e.g., acetonitrile, chloroform, dioxane, dichloromethane (DCM), dimethyl formamide (DMF), methyl chloride, 1,1,2-trichloroethane, or dichloroacetic acid, rather than water. The mixture of the solvents, such as DCM/DMF, chloroform/DMF and DCM/TFA was also reported [34-39]. In this case, polymer is dissolved in a suitable solvent and the desired number of fillers are also dispersed in the same solvent as the polymer and mixed, followed by the removal of the solvent, as shown in Figure 3.4. The solvent can be removed using distillation or lyophilisation. Yet the solution can be cast into the mould followed by solvent evaporation to afford the formation of the thin films. Some researcher reported on the use of melt-pressing as the last step to obtain thin composite films for characterization. This process ensures a complete solvent removal to afford smooth continuous films (see Figure 3.4).

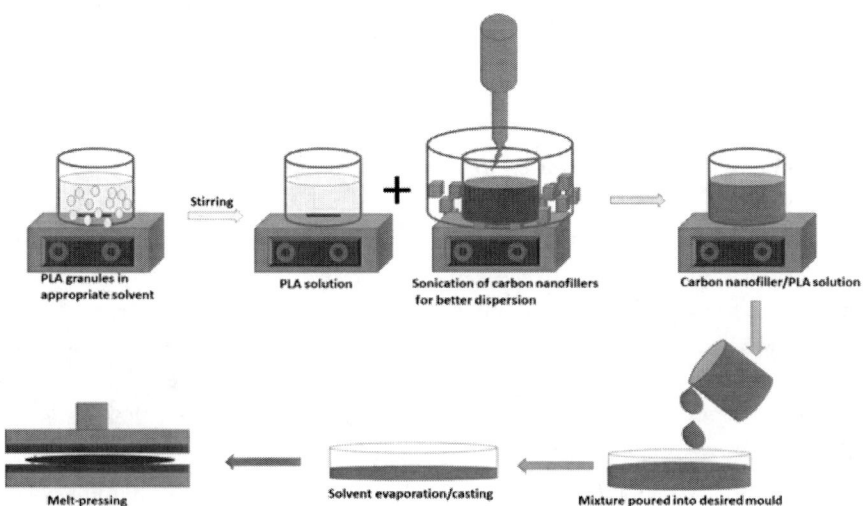

Figure 3.4. Schematic presentation of solution casting technique.

Table 3.2. Solution preparation of CBN-PLA composites

Formulation	Filler content (%)	Solvent used	Method	Highlights	Refs.
PLA/CNTs	1-7	THF	• Vacuum-dried • Melt-compression	Purification of CNTs resulted in strong interaction between PLA and CNTs which improved thermal, hardness, elastic modulus and surface resistivity	[40]
	0.5-10	Chloroform	• Vacuum-dried • Melt-compression	Good dispersion of CNTs with PLA which improved thermal properties, mechanical properties and electrical conductivity	[33]
	1-5		• Dried at 60°C for 12 hours • Vacuum-dried at 60°C for 6 hours	Treatment of the CNTs introduced functionalities while increasing available interface area that all promoted dispersion and interaction between PLA and CNTs	[41]
	0.05	CH$_2$Cl$_2$/toluene (2:1)	• Dried at room temperature for 24 hrs • Vacuum-dried at 40°C for 4 days • Vacuum-dried at 50°C for 150 minutes • Vacuum-dried at 60°C for 22 hrs	Homogeneous dispersion was attained that led to improved elastic modulus and tensile strength without affecting elongation-at-break	[38]
	0.5-2.5	DCM/THF 1:1	• Dried at room temperature for 24 hrs • Vacuum-dried at 60°C for 24 hrs	Purified CNTs were found to be homogeneously distributed with no visible aggregates Strong interfacial adhesion between components results from –COOH groups and –OH groups from PLA forming hydrogen bonds	[39]
	0.1-5	Chloroform	• Vacuum-dried at 80°C for 2 days • Melt-pressing at 180°C followed by quenching using ice	Grafting PLA onto CNTs promoted their dispersion within PLA	[42]
	1.2	Chloroform	• Oven-dried at 25°C for 24 hrs • Melt-pressed at 180°C	PLLA-grafting onto CNTs promoted dispersion and interaction with the host polymeric material, thus improved overall mechanical properties, and did not affect thermal conductivity	[43]

Table 3.2. (Continued).

Formulation	Filler content (%)	Solvent used	Method	Highlights	Refs.
	2%	Chloroform	• Oven-dried for 24 hrs • Melt-pressed at 180°C	The PLA-g-CNTs were wrapped around by PLA and exfoliated that allowed their dispersion without noticeable aggregates due to miscibility between PLA-g-CNTs and PLA	[34]
	0.1-3	THF	• Vacuum oven-dried • Melt-pressing	Despite the CNTs surface modification being used, the dimensional stability, storage modulus and electrical conductivity were improved without affecting thermal stability of PLA	[44]
	0.3-1	Dichloromethane	• Electrospinning	CNT aligned within the nanofibers was attained	[45]
	0.5	Dichloromethane	• Electrospinning	CNT aligned within the nanofibers was attained when electrodes' distance was kept at 3 cm	[46]
PLA/GP	0.4	Chloroform	• Vacuum-oven drying	Small number of graphene-based fillers are essential to improve mechanical properties to qualify for biomedical application	[47]
	0.5	Chloroform	• Vacuum-dried at 55°C for 2 days • Melt-pressed at 170°C	Good dispersion and interaction between PLA and GO were attained through surface modification using PLLA	[48]
	0.2	DMF	• Vacuum-dried at 80°C for 10 hrs	Intimate adhesion between PLA and graphene was achieved by freeze-drying graphene particles to afford their dispersion in polar organic solvent	[49]
	0.25-2	DMF	• Precipitated in water • Oven-dried at 60°C • Vacuum-dried at 60°C • Melt-pressed at 200°C for 3 minutes after preheating for 5 minutes	Good dispersion with strong interaction was attained which improved thermal stability and gas barrier properties of the ensuing composites	[50]
	1	DMF	• Vacuum-dried at 80°C for 4 days		[51]

Solution casting affords the dispersion of the neat CNTs through a covalent bond that was formed between the ester group in PLA and CNTs (Figure 3.5a) [52]. It can be seen that the surface was smooth without visible agglomerates. However, the modification of CNTs with a mixture of acids, viz. 1:3 HNO_3 and H_2SO_4, resulted in smoother surface nanocomposites with few visible pores, as shown in Figure 3.5b. This indicates that solution casting can improve the dispersion of the carbon nanofillers without the need for surface modification or the use of compatibilizers.

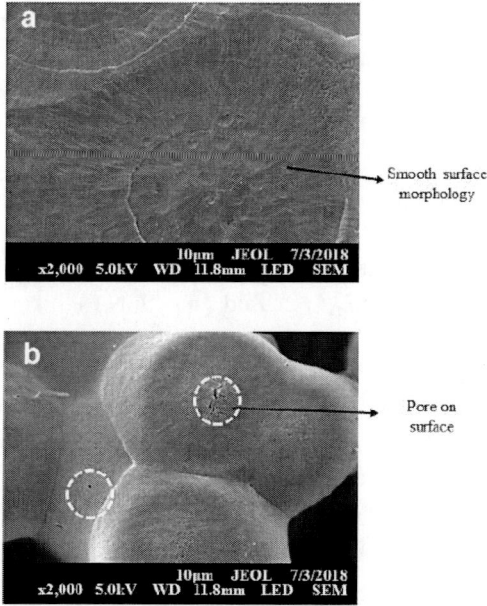

Figure 3.5. SEM images of (a) PLA/CNTs and (b) PLA/modified CNTs prepared through solution evaporation. Reproduced from Norazlina, et al. [52]. Open Access.

Some reports indicate that electrospinning can be used to produce fibres with diameters ranging between nano- and micro-meters [34, 36, 37, 53]. In this case, voltage is introduced to the solution, such that electrostatic repulsion overcomes the surface tension of the solution to produce jet travelling towards the collector. During this trip the solvent is evaporated to afford fibres [36, 37, 53]. The most commonly used solvent for the preparation of PLA/CNT composites is chloroform [33, 34, 38, 41-43, 54-55]. Other solvents such as THF [44], dichloromethane (DCM) [45], and the mixture of solvents (e.g., DMF/chloroform [36], DCM/THF [39]) were reported for the fabrication of PLA/CNT composites. McCullen et al. [36] prepared smooth electro-spun PLA/CNT composite nanofibers by dissolving PLA in a DMF/chloroform mixture. The fibres had diameters ranging between 550-960 nm, depending on the processing conditions (*viz.* voltage, tip-to-collector distance, and solvent used). The fibres obtained using chloroform alone resulted in larger diameters (i.e., 2.51 ± 2.13 μm) when compared to the fibres obtained from DMF/chloroform as a solvent (i.e., 0.55 ± 0.05 μm). In addition, the CNTs were found to be dispersed and aligned along the fibres with no sign of agglomeration. Elsewhere, the modification of CNTs using PCL and modified electrospinning technique promoted the alignment of the CNTs with the host polymeric material [45]. In the case of electrospinning, the electro-spun nanofibers were collected using grounded two parallel stainless steel collector to afford the alignment of the fibres, and thus the CNTs within the system. Using Raman spectroscopy, it was reported that there was strong Raman polarization, which indicates that CNTs were aligned within the fibres. Similar observations were reported by Kong et al. [46] using the same electrospinning set-up. They purified the CNTs using acid digestion followed by amino-functionalization using N,N'-dicyclohexyl carbodiimide (DCC) as condensation agent, and ethylenediamine (EDA). The amino-functionalized CNTs were then grafted with PCL to improve interfacial adhesion as well as the dispersion of CNTs because of structural similarities of PLA and PCL. The CNTs were well-embedded within the fibres and oriented along the axis of the fibres. However, SEM images showed that very few CNTs protruded onto the surface of the fibres, while most of these fillers were aligned with the fibres axis.

Moon et al. [33] prepared PLA/CNT composites using chloroform as a solvent, followed by sonication to improve the overall dispersion of the filler. The solvent was evaporated at room temperature by casting the solution onto the Teflon dishes. After a week, the as-prepared composites were vacuum-dried for 8 hours to ensure a complete removal of the solvent. Subsequently, the sample was melt-pressed at 200°C for 5 minutes to create 100-200 μm thin films. The CNTs were found to be well dispersed within the host matrix, which resulted in composites with good mechanical and electrical properties. Elsewhere, tetrahydrofuran (THF) was used as a solvent for the preparation of PLA/CNT composites [40]. The authors used the same procedure of drying the sample by placing it in a vacuum oven, followed by compression moulding. However, the authors compared the influence of purification of CNTs on their dispersion within PLA using solvent casting.

The purification of CNTs often introduces –COOH functionalities, which are essential for filler dispersion, especially when a highly polar solvent are used. In this case the CNTs were treated with a mixture of acids at a 50°C under sonication, followed by washing and drying. XRD results indicated that both diffraction peaks corresponded to PLA and amorphous carbon. It was noticed that there was no difference in the spectra of both purified and non-purified CNTs-based composites. However, the use of SEM demonstrated that the surface fracture for non-purified CNTs-based composites was smooth, with some holes resulting from poor interaction between PLA and CNTs. On the other hand, purified CNTs-based composites showed pull-outs of the filler with irregular shape surface fracture. This is due to strong interaction between PLA and purified CNTs because of the functionalities introduced during purification. The irregularities and pull-outs increased with an increase in CNTs content. It was, however, noticed that the average length of CNTs was slightly shorter when compared to the original CNTs, which could be due to damage during purification.

The functionalization of CNTs serves as one of the promoters to facilitate interaction and dispersion of CNTs within the host polymeric materials [38, 41]. Oxidative unzipping of CNTs was carried out by He et al. [41] by introducing CNTs into concentrated sulphuric acid, followed by the slow addition of $KMnO_4$. Such treatment introduces various surface functionalities such as carbonyl and carboxyl groups, but meanwhile opens and separates the walls of CNTs to enhance available interface area per volume to afford both inner and outer walls interacting with the host matrix. It was reported that the treatment of the CNTs improved their dispersion and interaction, which greatly enhanced the crystallization of PLA matrix and its hydrolytic degradation when compared to neat PLA. Covalent grafting of CNTs onto PLA was reported by Seligra, et al. [38]. This resulted in uniformly dispersed CNTs without visible agglomerations due to similarity in polarity. The latter resulted from surface functionalities introduced with strong interactions from the –COOH groups and C=O groups of PLA. However, the length of the CNTs was preserved.

Grafting of PLA onto CNTs was reported to promote interaction between PLA and CNTs [42, 43]. In addition, the dispersion is improved because of PLA chains wrapping around CNTs. The authors used chloroform as primary solvent for the preparation of PLA/CNT composites. The wrapping of PLA chains improved the overall mechanical properties of the composites; however, such morphology affect electrical conductivity negatively [42]. It was reported that unmodified CNTs-based composites exhibited superior electrical conductivity when compared to the modified ones [42]. Thus, the functionalization of the CNTs has to be carefully considered, depending on the anticipated application. The unmodified CNTs are essential when electrical conductivity is the primary requirement for the composites, while where mechanical performance is of significance, important functionalization can be adopted to afford better interaction and dispersion, which improves mechanical properties positively. Kim et al. [43] report that PLLA grafting via ring-opening polymerization (ROP) resulted in good mechanical properties. Tensile strength and modulus increased by ~13.1% and ~86.3% when PLLA-g-CNTs was introduced into PLA because of strong adhesion of CNTs onto PLA and the high miscibility between PLLA-g-CNTs and the host matrix. They found that elongation-at-break was improved; however, electrical conductivity for unmodified CNT-based composites was better than PLLA-g-CNT-based composites. This was attributed to PLLA hindering contact between CNTs. There was no significant difference for thermal conductivity for both grafted and unmodified CNTs-based composites. Yoon et al. [56] studied the effect of PLLA grafted onto CNTs on their dispersion within PLA using a solution casting method. In this case, the monomers were grafted via ROP onto carboxylic functionalized CNTs. It was noticed that the longer the grafted chain, the more interaction was attained between CNTs and PLA, which in turn facilitated the dispersion of the CNTs within the matrix. This demonstrates that the improvements in mechanical performance can be achieved through the modification of the carbon fillers using polymeric materials with similar properties as the host polymeric material. On the contrary, other applications that rely on the direct contact between fillers within the composite materials could be negatively affected. For instance, electrical conductivity is often lower for modified fillers, despite better dispersion and interfacial adhesion when compared to unmodified filler-based composites, because the contact between the fillers will be hindered by a surface-coated layer of the polymeric material. Therefore, the surface modification has to be considered carefully to avoid polymeric material wrapping around the carbon-based filler to avoid a negative impact on other properties rather than mechanical properties. In their work, Chiu et al. [44] demonstrate that the modification of CNTs can be achieved without affecting the electrical conductivity of PLA/CNT composites negatively. The authors used purified CNTs with a mixture of sulphuric and nitric acids, followed by grafting 3-isocyanatopropyl triethoxysilane (IPTES) to afford highly dispersed CNTs with strong interfacial interaction between CNTs and PLA. Purification of the CNTs was carried out by using a combination of sonication and microwave-assisted acid digestion in order to avoid damages on the CNTs while removing impurities at less solvent consumption. No pull-outs were present when modified CNTs were used as fillers, regardless of the CNTs content. This was ascribed to hydrogen bonds formed between modified CNTs and PLA. It was reported that electrical conductivity increased with an increase in concentration of modified CNTs within the system.

Cao et al. [49] used a lyophilization method to improve the dispersion of graphene nanofillers in organic polar solvents which can dissolve PLA. It was found that the resulting lyophilized nanofillers can be redispersed in DMF to mix easily with PLA. The mixture was agitated for 2 hours at 85°C, sonicated for 2 hours at 70°C and then coagulated in methanol. The as prepared mixture was filtered and then vacuum dried for 10 hours at 80°C to afford well-dispersed fillers with strong interfacial adhesion between the composite components. This offers a cheaper method to avoid the use of expensive chemicals and a complex process to modifying either PLA and/or carbon fillers to improve dispersion and interaction between the composite components. The main issue is the appropriate solvents, which could be expensive, especially for industrial-scale production.

Similar to CNT-based composites, chloroform is the most utilized solvent for the preparation of graphene-based composites via solution casting [55, 47, 48, 57]. Other solvents, including DMF [49, 50, 51, 58-60] as well as co-solvents were also reported for such purpose. In their work, Pinto et al. [57] used chloroform as suitable solvent to prepare PLA/GP composites. They reported that both graphene oxide (GO) and graphene nanoplatelets (GNP) were homogenously dispersed within PLA with some visible aggregates. These resulted in good mechanical performance and barrier properties. Elsewhere, graphene nanosheets (GNS) were introduced into PLA. In this regard, GNS were dispersed in DMF using ultrasonication for 2 hours and thoroughly mixed with PLA solution by stirring for 4 hours at 40°C to ensure complete homogeneity. The mixture was then sonicated for 10 minutes prior to casting. It was noticed that the GNS were randomly distributed through the matrix (presented by whites spots) (Figure 3.6a) with some filler pull-outs and irregular holes (indicated by arrows) (Figure 3.6b). However, no exfoliation of the graphene sheets was observed, with most of them maintaining their natural, multilayered structure (Figure 3.6c). This shows that a lack of strong shear in solution mixing limits the intercalation between PLA chains and the nanosheets to some extent. Some of the single-layered and multi-layered fragments were observed, indicating that the resulting product was mixture of a micro- and nano-composite (Figure 3.6d).

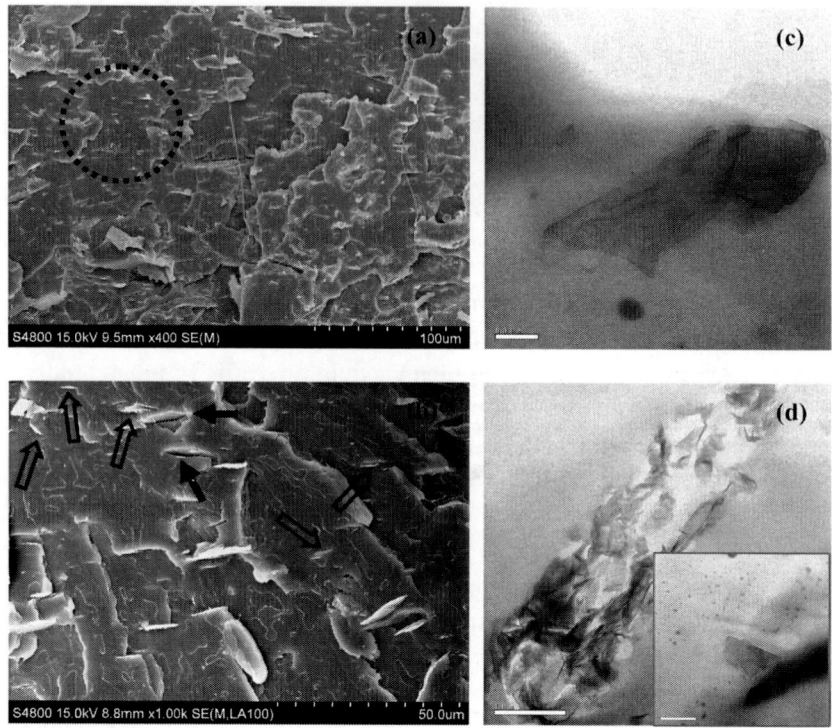

Figure 3.6. SEM images of PLA/GNS composite. (b) Enlarged image of the circled are in panel a. (c-d) TEM images of PLA/GNS composite. Adapted with permission from Wu et al.[51]. Copyright 2013 American Chemical Society.

The main issue associated with graphene-based nanocomposites is their irreversible aggregation. A number of researchers reported on the use of surface modification to overcome this issue [55, 48]. Li et al. [48] grafted PLLA onto GO through melt-polycondensation process to improve its dispersion within the PLA matrix. In comparison to unmodified GO, PLLA-g-GO-based composites displayed crude surface area, indicating that more energy was required to break the sample because of the strong interfacial adhesion between PLA and PLLA-g-GO. For PLA/GO composites, clusters of GO were visible, resulting from irreversible aggregation resulting from

Van der Waal's forces among GO, thus hindering interaction with the host polymeric material. For PLLA-g-GO-based composite no visible aggregates were observed, because the grafted PLLA was wrapped around GO promoting dispersion and interfacial adhesion between the components. This was reflected by an increase in breaking strength and distortion at break by ~114% and ~197%, respectively, for PLLA-g-GO/PLA composites with PLA/GO exhibiting an increase by 95% and 144% when compared to neat PLA. Surface modification of exfoliated graphene (EG) with silane resulted in composites, with the same mechanical performance as those based on CNTs [55]. These composites outperformed CNT-based composites when it comes to thermomechanical properties. The resulting performance was associated with strong interfacial adhesion as well as good filler dispersion. It is worth noting that the increase in graphene nanofillers resulted in more aggregates being observed in the system [59]. Wang et al. [59] report that when low graphene content (0.5%) is employed as reinforcing phase there is no visible clusters, but as the content increases (1-2%), some aggregates are observed. This demonstrates that there is optimal fillers' content at which good dispersion and strong interfacial between the filler and host polymeric material can be achieved, and thus improvements on the overall properties of the resulting PLA/GN composites.

In their work, Huang et al. [50] report on the modification of the conventional solvent casting method by precipitating the PLA/graphene oxide nanosheets (GONS) mixture in water followed by melt-pressing. The main objective was to exploit graphene oxide having a large number of oxygen-containing functional groups (e.g., hydroxyl, epoxide, carbonyl and carboxyl) on the basal and edge planes to improve interaction by forming hydrogen bonding with the available ester groups on PLA, meanwhile overcoming filler sedimentation during solvent evaporation. In this case, the authors melt-pressed the precipitated mixture for 3 minutes after preheating for 5 minutes to ensure that the exfoliation and random distribution of the filler are preserved. GONS were individually exfoliated (indicated by arrows) and randomly distributed with the host matrix because of weak flow field applied during the melt-pressing process. The rough surface fracture of the nanocomposite was attributed to the filler being well-embedded in PLA and strong interfacial adhesion resulting from hydrogen bonding from oxygen-containing GONS and PLA ester groups. However, neat PLA showed a smooth fractured surface. In addition, the wrinkled GONS structure facilitates mechanical interlocking with PLA.

An electrospinning technique has been employed for the preparation of graphene/PLA composites [61]. In this regard the dispersion of the filler is improved because the fibres are in nano- to micro-scale [61]. Sisti et al. [61] prepared electro-spun nanofibers of PLA containing graphene using DMF as solvent. Smooth, bead-free nanofibers with smaller diameters, i.e., 250 ±40 nm, than neat PLA-based nanofibers with a diameter of 300 ± 50 nm. The reduction in the diameters is because of an increase in solution conductivity. It was reported that the graphene sheets were stacked and aggregated inside the fibres. Some of the nanosheets were completely spread along the fibres.

Su et al. [62] prepared PLA/CB composites using chloroform as solvent of choice. The authors compared the effect of modified and unmodified CB on the crystallization behaviour of PLA. The modification of CB was carried out using an internal melt-mixer by mixing CB with 3,9-bis[1,1-dimethyl-2(-(3-tert-butyl-4-hydroxyl-5-methylphenyl)propionyloxy)ethyl]−2,4,8,10-tertaoxaspiro-[5,5]-undecane (AO-80) at 1/08 ratio at 140°C for ~30 minutes. Besides the fact that both unmodified CB and modified CB (MCB) act as good nucleating agent for PLA, MCB had better nucleation activity than neat CB due to strong interfacial adhesion.

3.2.4. *In Situ* Polymerization

In situ polymerization involves the polymerization of PLA monomers in the presence of carbon nanofillers (Table 3.3) [63-66]. In most cases the carbon fillers have to be modified to be furnished with functionalities that can assist in anchoring the growth of polymer chains from the employed monomers [63].

The modification of CNTs has been the first step to introduce functionalities that can initiate the in-situ polymerization process from PLA monomers [63, 65-68]. As mentioned earlier, the mixture of nitric acid and sulphuric acid is often used to treat CNTs to afford the introduction of carboxyl, hydroxyl, aldehyde, or ketone groups, which are essential for surface-initiated polymerization [63, 67]. In their work, Chen et al. [67] introduced hydroxyl groups onto CNTs to serve as co-initiators for polymerization of L-lactide via surface-initiated ring-opening polymerization. The authors utilized two solvents, i.e., DMF and toluene at 140°C and 70°C, respectively. The content of PLA within the composite material increased from 10-35% with an increase in polymerization time from 2 to 20 hours in DMF, and from 7-32% in toluene. It was reported that

PLLA had strong interaction with CNTs with good dispersion. To further improve the interaction and dispersibility of CNTs, modification is often applied onto the purified CNTs. For instance, Boncel et al. [63] purified CNTs with a mixture of acid to afford a large number of carboxyl groups (COOH-CNTs) to afford linkage with the linker, *viz. β-D*-uridine. The authors reacted the COOH-CNTs with $SOCl_2$ to remove the –OH with Cl groups to afford modification with the linker. The functionalized CNTs/PLA composites were prepared using (*R,R*)-lactide monomer for polymerization in the presence of stannous octanoate ($Sn(Oct)_2$) and CNTs as catalysts and/or initiators. The particles were well-distributed within the PLA matrix. Elsewhere, it was reported that Bergman cyclization can be employed for surface modification of the CNTs to allow grafting polylactide [68]. These particles serve as initiators to afford attachment of PLA via ring-opening polymerization.

Table 3.3. In situ polymerization PLA/carbon fillers composites

Formulation	Method	Catalyst	Conditions	Highlights	Refs.
PLA/CNT	Grafting from	CNT-β-D-uridine/ $Sn(Oct)_2$	DMF, 130°C, 24 hrs	Better CNTs dispersion at ~7% with PLA molecular weight of 116 700 Da were attained	[63]
PLA/CNT	Grafting from	CNT-Titanium alkoxide	Toluene 130°C for 20 hrs	The amount of grafted PLA can be controlled with time	[69]
	Grafting to	Erythritol/SN$(Oct)_2$/ CNTs-COCl	Toluene, 110°C, 48 hrs	Five-armed PLA was covalently bonded to CNTs which promoted its dispersion for preparation of composites	[65]
	Grafting to	CNT-pyrrolidine groups	DMF, 135°C, 3 hrs	Covalent bonding of PLA in the melt was achieved	[66]
PLA/GO	Grafting from	GO, $SN(Oct)_2$	170°C, 4 hrs, chloroform	There was no correlation between the content of the GO and molecular weight of PLA attained	[70]
PLA/GO	Grafting from	GO	120°C, 12 hrs, water/toluene		[71]

Feng et al. [64] report on the use of grafting PLA onto the surface of magnetic-CNT (m-CNTs) through in situ ring-opening polymerization of lactide. It was noticed that the amount of PLLA grafted onto the m-CNTs can be controlled by varying the feeding ratio of the monomer to m-CNTs. The m-CNTs were obtained by in situ, high temperature decomposition using iron (III) acetylacetonate as precursor in polyol solution [72]. The nanoparticles' size and their coverage on CNTs could be adjusted by changing the ratio of precursor to CNTs. SEM and TEM images confirmed that the PLLA was successfully grafted onto m-CNTs with the polymer thickness of 2.5 nm covering the CNTs. Besides the fact that purified CNTs and stannous acetate often employed an initiator and/or catalyst for the polymerization L-lactide, some reports demonstrated that other catalysts can be used for the same purpose [69]. Priftis et al. [69] used CNTs functionalized with titanium alkoxide via a Diels-Alder cycloaddition for coordination polymerization of L-lactide. A polymer layer of ~23 nm on the surface of the CNTs was obtained. It was demonstrated that by varying time, the amount of PLA grown on top of the functionalized CNTs can be controlled.

A number of studies reported on the in-situ polymerization of L-lactide in the presence of graphene-based nanofillers [70, 71, 73]. Wang et al. [71] prepared graphene oxide (GO) from natural graphite through a conventional Hummers method to afford the introduction of hydroxyl and carboxyl groups that are essential for polymerization of L-lactide. It was reported that L-lactic acid was able to graft onto GO via reaction between the available –OH and –COOH. It was also noticed that while condensation of PLA monomers takes place at higher temperatures, that facilitates the polymerization of PLA to afford high molecular weight PLA grafted onto GO. In their study, Yang et al. [70] thermally reduced graphene oxide obtained from natural graphite using traditional Hummer's method. In this regard, GO was introduced into a tube and purged with inert gas for 5 minutes and subsequently introduced into a 1000°C furnace for a minute. The resulting thermally reduced GO (TRGO) had less intense peaks associated with –OH, –COOH, and carbonyl groups, indicating that the thermal reduction of GO was successful. TRGO served as initiator with tin octoate as the catalyst for in-situ ring-opening polymerization L-lactide. SEM and TEM images confirmed that there was strong interaction between TRGO and PLA, which resulted in good dispersion of the graphene nanosheets within the host polymeric material. The modification of graphene-based fillers to improve the grafting of PLA onto the fillers was studied by Pramoda et al. [73]. The authors functionalized GO with tolylene-2, 4-diisocyanate (TDI) in the presence of 1,4-butanediol to afford the introduction of a large number –OH groups and compared with GO-graft- polyhedral oligomeric silsesquioxane (POSS), i.e.,

POSS 8 hydroxyl group grafted onto GO nanosheets. FTIR results confirmed that all OH groups from functionalized GO were completely used up during in-situ polymerization of PLA. However, for unmodified GO not all –OH were consumed during polymerization, demonstrating that modification of GO is essential to facilitate the polymerization of PLA. In the case of POSS-g-GO, a complete consumption of –OH groups was achieved indicating the importance of GO modification. From NMR results, it was deduced that POSS-g-GO exhibited composites with a higher molecular weight value of ~10300 g mol^{-1}; meanwhile functionalized GO-based composite had a molecular weight of ~6100g mol^{-1}.

3.3. Conclusion

Solution mixing is the most commonly reported preparation technique for PLA/CBF composites when compared to other preparation methods, i.e., melt-compounding and in situ polymerization. Besides being the least reported preparation method, CBF/PLA composites prepared via in situ polymerization afford good dispersion of the carbon fillers with a strong interfacial adhesion essential for the overall resulting properties. The industrial scale-up of this method is the primary concern. Melt-intercalation as the second-most employed method suffers from few disadvantages. One of the main issues is the amount of the CBF required. In addition, the irreversible aggregation CBFs because of the Van der Waals forces could drastically affect the interfacial adhesion, and thus the resulting properties. Therefore, the modification of PLA or CBF is mandatory to avoid these issues. Nevertheless, melt-compounding is the most attractive technique because of its eco-friendliness and industrial scalability.

In general, the use of unmodified CBF as filler for PLA results in their inhomogeneous distribution, regardless of the preparation method used. The surface modification of CBFs from purification to grafting of chemical substances in order to afford better dispersion as well as strong interaction between the composite components has been the main focal subject. These modifications, however, can affect the properties of the resulting composite materials negatively. For instance, grafting of PLA onto CNTs is the most commonly used technique that improves dispersion and interfacial adhesion due to high miscibility between PLA-g-CNTs and PLA host matrix. However, the resulting composite is usually recognized by poor electrical conductivity due to the thin layer covering the filler; hence breaking the conduction path within the resulting composite. On the other hand, these composites were found to exhibit excellent mechanical performance, remarkable thermomechanical and elevated thermal stability. On the other hand, the content of the filler is another important parameter. The optimal content of CBF is required to achieve the desired properties for the anticipated application. This demonstrates that the chemical modification and filler content are crucial parameters after the preparation method to finetune the properties of the composites towards anticipated application.

References

[1] Lim LT, Auras R, Rubino M. Processing technologies for poly(lactic acid). *Progress in Polymer Science* 2008; 33:820-52.
[2] Mokhena TC, Sefadi JS, Sadiku ER, John MJ, Mochane MJ, Mtibe A. Thermoplastic Processing of PLA/Cellulose Nanomaterials Composites. *Polymers* 2018;10.
[3] Mofokeng JP, Luyt AS. Morphology and thermal degradation studies of melt-mixed poly(lactic acid) (PLA)/poly(ε-caprolactone) (PCL) biodegradable polymer blend nanocomposites with TiO_2 as filler. *Polymer Testing* 2015; 45:93-100.
[4] Mokhena TC, Mochane MJ, Sadiku ER, Agboola O, John MJ. Opportunities for PLA and Its Blends in Various Applications. In: Gnanasekaran D, editor. *Green Biopolymers and their Nanocomposites.* Singapore: Springer Singapore; 2019. p. 55-81.
[5] Mofokeng JP, Luyt AS. Morphology and thermal degradation studies of melt-mixed poly(hydroxybutyrate-co-valerate) (PHBV)/poly(ε-caprolactone) (PCL) biodegradable polymer blend nanocomposites with TiO_2 as filler. *Journal of Materials Science* 2015; 50:3812-24.
[6] Mokhena TC, Jacobs V, Luyt A. *A review on electrospun bio-based polymers for water treatment.* 2015.
[7] Mochane MJ, Mokhena TC, Sadiku ER, Ray SS, Mofokeng TG. Green Polymer Composites Based on Polylactic Acid (PLA) and Fibers. In: Gnanasekaran D, editor. *Green Biopolymers and their Nanocomposites.* Singapore: Springer Singapore; 2019. p. 29-54.
[8] Liu Y-L. Effective approaches for the preparation of organo-modified multi-walled carbon nanotubes and the corresponding MWCNT/polymer nanocomposites. *Polymer Journal* 2016; 48:351-8.
[9] Lin W-Y, Shih Y-F, Lin C-H, Lee C-C, Yu Y-H. The preparation of multi-walled carbon nanotube/poly(lactic acid) composites with excellent conductivity. *Journal of the Taiwan Institute of Chemical Engineers* 2013; 44:489-96.
[10] Raja M, Ryu SH, Shanmugharaj AM. Thermal, mechanical and electroactive shape memory properties of polyurethane (PU)/poly (lactic acid) (PLA)/CNT nanocomposites. *European Polymer Journal* 2013; 49:3492-500.

[11]   Pötschke P, Andres T, Villmow T, Pegel S, Brünig H, Kobashi K, et al. Liquid sensing properties of fibres prepared by melt-spinning from poly(lactic acid) containing multi-walled carbon nanotubes. *Composites Science and Technology* 2010; 70:343-9.

[12]   Wu C-S, Liao H-T. Study on the preparation and characterization of biodegradable polylactide/multi-walled carbon nanotubes nanocomposites. *Polymer* 2007; 48:4449-58.

[13]   Masiuchok O, Iurzhenko M, Kolisnyk R, Mamunya Y, Godzierz M, Demchenko V, et al. Polylactide/Carbon Black Segregated Composites for 3D Printing of Conductive Products. *Polymers* 2022.

[14]   Liang L, Huang T, Yu S, Cao W, Xu T. Study on 3D printed graphene/carbon fiber multi-scale reinforced PLA composites. *Materials Letters* 2021; 300:130173.

[15]   Guo J, Tsou C-H, Yu Y, Wu C-S, Zhang X, Chen Z, et al. Conductivity and mechanical properties of carbon black-reinforced poly(lactic acid) (PLA/CB) composites. *Iranian Polymer Journal* 2021; 30:1251-62.

[16]   Wang N, Zhang X, Ma X, Fang J. Influence of carbon black on the properties of plasticized poly(lactic acid) composites. *Polymer Degradation and Stability* 2008; 93:1044-52.

[17]   Yu J, Wang N, Ma X. Fabrication and Characterization of Poly(lactic acid)/Acetyl Tributyl Citrate/Carbon Black as Conductive Polymer Composites. *Biomacromolecules* 2008; 9:1050-7.

[18]   Villmow T, Pötschke P, Pegel S, Häussler L, Kretzschmar B. Influence of twin-screw extrusion conditions on the dispersion of multi-walled carbon nanotubes in a poly(lactic acid) matrix. *Polymer* 2008; 49:3500-9.

[19]   Cardoso PHM, de Oliveira MFL, de Oliveira MG, da Silva Moreira Thiré RM. 3D Printed Parts of Polylactic Acid Reinforced with Carbon Black and Alumina Nanofillers for Tribological Applications. *Macromolecular Symposia* 2020; 394:2000155.

[20]   Delgado PA, Brutman JP, Masica K, Molde J, Wood B, Hillmyer MA. High surface area carbon black (BP-2000) as a reinforcing agent for poly[(−)-lactide]. *Journal of Applied Polymer Science* 2016;133.

[21]   Bortoli LSD, Farias Rd, Mezalira DZ, Schabbach LM, Fredel MC. Functionalized carbon nanotubes for 3D-printed PLA-nanocomposites: Effects on thermal and mechanical properties. *Materials Today Communications* 2022; 31:103402.

[22]   Yang L, Li S, Zhou X, Liu J, Li Y, Yang M, et al. Effects of carbon nanotube on the thermal, mechanical, and electrical properties of PLA/CNT printed parts in the FDM process. *Synthetic Metals* 2019; 253:122-30.

[23]   Valvez S, Santos P, Parente JM, Silva MP, Reis PNB. 3D printed continuous carbon fiber reinforced PLA composites: A short review. *Procedia Structural Integrity* 2020; 25:394-9.

[24]   Zhou N, Yang S, Liu Y, Tuo X, Gong Y, Guo J. Performance evaluation on particle-reinforced rigid/flexible composites via fused deposition modeling 3D printing. *Journal of Applied Polymer Science* 2022; 139:52149.

[25]   Esmizadeh E, Sadeghi T, Vahidifar A, Naderi G, Ghoreishy MHR, Paran SMR. Nano Graphene-Reinforced Bio-nanocomposites Based on NR/PLA: The

Morphological, Thermal and Rheological Perspective. *Journal of Polymers and the Environment* 2019; 27:1529-41.
[26] Marconi S, Alaimo G, Mauri V, Torre M, Auricchio F. Impact of graphene reinforcement on mechanical properties of PLA 3D printed materials. *2017 IEEE MTT-S International Microwave Workshop Series on Advanced Materials and Processes for RF and THz Applications* (IMWS-AMP)2017. p. 1-3.
[27] Maqsood N, Rimašauskas M. Characterization of carbon fiber reinforced PLA composites manufactured by fused deposition modeling. *Composites Part C: Open Access* 2021; 4:100112.
[28] Tian X, Liu T, Yang C, Wang Q, Li D. Interface and performance of 3D printed continuous carbon fiber reinforced PLA composites. *Composites Part A: Applied Science and Manufacturing* 2016; 88:198-205.
[29] Sathies T, Senthil P, Prakash C. Application of 3D printed PLA-carbon black conductive polymer composite in solvent sensing. *Materials Research Express* 2019; 6:115349.
[30] Schmack G, Tändler B, Optiz G, Vogel R, Komber H, Häußler L, et al. High-speed melt-spinning of various grades of polylactides. *Journal of Applied Polymer Science* 2004; 91:800-6.
[31] Ghosh S, Vasanthan N. Structure development of poly(L-lactic acid) fibers processed at various spinning conditions. *Journal of Applied Polymer Science* 2006; 101:1210-6.
[32] Schmack G, Tändler B, Vogel R, Beyreuther R, Jacobsen S, Fritz H-G. Biodegradable fibers of poly(L-lactide) produced by high-speed melt spinning and spin drawing. *Journal of Applied Polymer Science* 1999; 73:2785-97.
[33] Moon S-I, Jin F, Lee C-j, Tsutsumi S, Hyon S-H. novel carbon nanotube/poly(L-lactic acid) Nanocomposites; Their Modulus, Thermal Stability, and Electrical Conductivity. *Macromolecular Symposia* 2005; 224:287-96.
[34] Kim H-S, Hyun Park B, Yoon J-S, Jin H-J. Thermal and electrical properties of poly(l-lactide)-graft-multiwalled carbon nanotube composites. *European Polymer Journal* 2007; 43:1729-35.
[35] Luo J, Wang H, Zuo D, Ji A, Liu Y. Research on the Application of MWCNTs/PLA Composite Material in the Manufacturing of Conductive Composite Products in 3D Printing. *Micromachines* 2018.
[36] McCullen SD, Stano KL, Stevens DR, Roberts WA, Monteiro-Riviere NA, Clarke LI, et al. Development, optimization, and characterization of electrospun poly(lactic acid) nanofibers containing multi-walled carbon nanotubes. *Journal of Applied Polymer Science* 2007; 105:1668-78.
[37] Zeng J, Xu X, Chen X, Liang Q, Bian X, Yang L, et al. Biodegradable electrospun fibers for drug delivery. *Journal of Controlled Release* 2003; 92:227-31.
[38] Seligra PG, Nuevo F, Lamanna M, Famá L. Covalent grafting of carbon nanotubes to PLA in order to improve compatibility. *Composites Part B: Engineering* 2013; 46:61-8.
[39] Chrissafis K, Paraskevopoulos KM, Jannakoudakis A, Beslikas T, Bikiaris D. Oxidized multiwalled carbon nanotubes as effective reinforcement and thermal

stability agents of poly(lactic acid) ligaments. *Journal of Applied Polymer Science* 2010; 118:2712-21.

[40] Chiu W-M, Chang Y-A, Kuo H-Y, Lin M-H, Wen H-C. A study of carbon nanotubes/biodegradable plastic polylactic acid composites. *Journal of Applied Polymer Science* 2008; 108:3024-30.

[41] He L, Sun J, Wang X, Fan X, Zhao Q, Cai L, et al. Unzipped multiwalled carbon nanotubes-incorporated poly(l-lactide) nanocomposites with enhanced interface and hydrolytic degradation. *Materials Chemistry and Physics* 2012; 134:1059-66.

[42] Yoon JT, Jeong YG, Lee SC, Min BG. Influences of poly(lactic acid)-grafted carbon nanotube on thermal, mechanical, and electrical properties of poly(lactic acid). *Polymers for Advanced Technologies* 2009; 20:631-8.

[43] Kim H-S, Chae YS, Park BH, Yoon J-S, Kang M, Jin H-J. Thermal and electrical conductivity of poly(l-lactide)/multiwalled carbon nanotube nanocomposites. *Current Applied Physics* 2008; 8:803-6.

[44] Chiu W-M, Kuo H-Y, Tsai P-A, Wu J-H. Preparation and Properties of Poly (Lactic Acid) Nanocomposites Filled with Functionalized Single-Walled Carbon Nanotubes. *Journal of Polymers and the Environment* 2013; 21:350-8.

[45] Kong Y, Yuan J, Wang Z, Qiu J. Study on the preparation and properties of aligned carbon nanotubes/polylactide composite fibers. *Polymer Composites* 2012; 33:1613-9.

[46] Kong Y, Yuan J, Qiu J. Preparation and characterization of aligned carbon nanotubes/polylactic acid composite fibers. *Physica B: Condensed Matter* 2012; 407:2451-7.

[47] Pinto AM, Moreira S, Gonçalves IC, Gama FM, Mendes AM, Magalhães FD. Biocompatibility of poly(lactic acid) with incorporated graphene-based materials. *Colloids and Surfaces B: Biointerfaces* 2013; 104:229-38.

[48] Li W, Xu Z, Chen L, Shan M, Tian X, Yang C, et al. A facile method to produce graphene oxide-g-poly(L-lactic acid) as an promising reinforcement for PLLA nanocomposites. *Chemical Engineering Journal* 2014; 237:291-9.

[49] Cao Y, Feng J, Wu P. Preparation of organically dispersible graphene nanosheet powders through a lyophilization method and their poly(lactic acid) composites. *Carbon* 2010; 48:3834-9.

[50] Huang H-D, Ren P-G, Xu J-Z, Xu L, Zhong G-J, Hsiao BS, et al. Improved barrier properties of poly(lactic acid) with randomly dispersed graphene oxide nanosheets. *Journal of Membrane Science* 2014; 464:110-8.

[51] Wu D, Cheng Y, Feng S, Yao Z, Zhang M. Crystallization Behavior of Polylactide/Graphene Composites. *Industrial & Engineering Chemistry Research* 2013; 52:6731-9.

[52] Norazlina H, Suhaila A, Nabihah A, Rabiatul MM, Zulhelmie I, Yusoh Y. Degradation behaviour of plasticized PLA/CNTs nanocomposites prepared by the different technique of blending. *IOP Conference Series: Materials Science and Engineering* 2021; 1068:012002.

[53] Zeng J, Chen X, Xu X, Liang Q, Bian X, Yang L, et al. Ultrafine fibers electrospun from biodegradable polymers. *Journal of Applied Polymer Science* 2003; 89:1085-92.

[54] Gonçalves C, Gonçalves IC, Magalhães FD, Pinto AM. Poly(lactic acid) Composites Containing Carbon-Based Nanomaterials: A Review. *Polymers* 2017; 9:269.

[55] Li W, Shi C, Shan M, Guo Q, Xu Z, Wang Z, et al. Influence of silanized low-dimensional carbon nanofillers on mechanical, thermomechanical, and crystallization behaviors of poly(L-lactic acid) composites—A comparative study. *Journal of Applied Polymer Science* 2013; 130:1194-202.

[56] Yoon JT, Lee SC, Jeong YG. Effects of grafted chain length on mechanical and electrical properties of nanocomposites containing polylactide-grafted carbon nanotubes. *Composites Science and Technology* 2010; 70:776-82.

[57] Pinto AM, Cabral J, Tanaka DAP, Mendes AM, Magalhães FD. Effect of incorporation of graphene oxide and graphene nanoplatelets on mechanical and gas permeability properties of poly(lactic acid) films. *Polymer International* 2013; 62:33-40.

[58] Wang H, Qiu Z. Crystallization behaviors of biodegradable poly(l-lactic acid)/graphene oxide nanocomposites from the amorphous state. *Thermochimica Acta* 2011; 526:229-36.

[59] Wang H, Qiu Z. Crystallization kinetics and morphology of biodegradable poly(l-lactic acid)/graphene oxide nanocomposites: Influences of graphene oxide loading and crystallization temperature. *Thermochimica Acta* 2012; 527:40-6.

[60] Shen Y, Jing T, Ren W, Zhang J, Jiang Z-G, Yu Z-Z, et al. Chemical and thermal reduction of graphene oxide and its electrically conductive polylactic acid nanocomposites. *Composites Science and Technology* 2012; 72:1430-5.

[61] Sisti L, Belcari J, Mazzocchetti L, Totaro G, Vannini M, Giorgini L, et al. Multicomponent reinforcing system for poly(butylene succinate): Composites containing poly(l-lactide) electrospun mats loaded with graphene. *Polymer Testing* 2016; 50:283-91.

[62] Su Z, Li Q, Liu Y, Guo W, Wu C. The nucleation effect of modified carbon black on crystallization of poly(lactic acid). *Polymer Engineering & Science* 2010; 50:1658-66.

[63] Boncel S, Brzeziński M, Mrowiec-Białoń J, Janas D, Koziol KKK, Walczak KZ. Oxidised multi-wall carbon nanotubes–(R)-polylactide composite with a covalent β-d-uridine filler-matrix linker. *Materials Letters* 2013; 91:50-4.

[64] Feng J, Cai W, Sui J, Li Z, Wan J, Chakoli AN. Poly(l-lactide) brushes on magnetic multiwalled carbon nanotubes by in situ ring-opening polymerization. *Polymer* 2008; 49:4989-94.

[65] Xu Z, Niu Y, Yang L, Xie W, Li H, Gan Z, et al. Morphology, rheology and crystallization behavior of polylactide composites prepared through addition of five-armed star polylactide grafted multiwalled carbon nanotubes. *Polymer* 2010; 51:730-7.

[66] Novais RM, Simon F, Pötschke P, Villmow T, Covas JA, Paiva MC. Poly(lactic acid) composites with poly(lactic acid)-modified carbon nanotubes. *Journal of Polymer Science Part A: Polymer Chemistry* 2013; 51:3740-50.

[67] Chen G-X, Kim H-S, Park BH, Yoon J-S. Synthesis of Poly(L-lactide)-Functionalized Multiwalled Carbon Nanotubes by Ring-Opening Polymerization. *Macromolecular Chemistry and Physics* 2007; 208:389-98.

[68] Ma J, Cheng X, Ma X, Deng S, Hu A. Functionalization of multiwalled carbon nanotubes with polyesters via bergman cyclization and "grafting from" strategy. *Journal of Polymer Science Part A: Polymer Chemistry* 2010; 48:5541-8.

[69] Priftis D, Petzetakis N, Sakellariou G, Pitsikalis M, Baskaran D, Mays JW, et al. Surface-Initiated Titanium-Mediated Coordination Polymerization from Catalyst-Functionalized Single and Multiwalled Carbon Nanotubes. *Macromolecules* 2009;42: 3340-6.

[70] Yang J-H, Lin S-H, Lee Y-D. Preparation and characterization of poly(l-lactide)–graphene composites using the in situ ring-opening polymerization of PLLA with graphene as the initiator. *Journal of Materials Chemistry* 2012; 22:10805-15.

[71] Wang L-N, Guo Wang P-Y, Wei J-C. Graphene Oxide-Graft-Poly(l-lactide)/Poly(l-lactide) Nanocomposites: Mechanical and Thermal Properties. *Polymers* 2017.

[72] Wan J, Cai W, Feng J, Meng X, Liu E. In situ decoration of carbon nanotubes with nearly monodisperse magnetite nanoparticles in liquid polyols. *Journal of Materials Chemistry* 2007; 17:1188-92.

[73] Pramoda KP, Koh CB, Hazrat H, He CB. Performance enhancement of polylactide by nanoblending with POSS and graphene oxide. Polymer Composites 2014; 35:118-26.

Chapter 4

Thermal Stability of PLA/Carbon Based Filler(s) Composites

Abstract

When polymeric materials are heated, they undergo both a physical and chemical transition. There is a need for polymeric materials, including PLA, to be used in high-temperature applications. Based on the above statement, there is a need to improve the thermal stability of the PLA. Carbon-based fillers have the ability to be resistant to heat; as a result, they have the ability of enhancing the thermal stability of the PLA matrix. More stable PLA-based composites has huge potential to be used in engineering applications. This chapter discusses the thermal stability of PLA/carbon-based fillers.

Keywords: thermal stability, graphene, hybridization, functionalization, heat

4.1. Introduction

PLA is a semicrystalline plastic from the aliphatic polyester family [1-4]. It is derived from natural resources such as wheat, rice and beetroot. It is widely used in food (packaging and tableware) and healthcare (tissue engineering and drug delivery) industries because of its biocompatibility, biodegradability, and sustainability [1-5]. It is inherently hydrophobic because of the methyl side groups (see Figure 4.1). Besides the fact that PLA is hydrophobic polymers, it can easily degrade under hydrolytic conditions with water molecules breaking the ester bonds. Such chain scission results in significant molecular weight reductions and affects mechanical properties drastically. Thermal stability of PLA and PLA-based composites is an essential parameter to evaluate the extent of degradation before and after processing. In most cases, the first step in preparing PLA and PLA-based composites involves the drying process to avoid moisture absorption of the samples, which can lead to material degradation [3]. The insight into the extent of PLA degradation is crucial for its long-term use and essentially informs end-of-life strategies, namely recovery and recyclability.

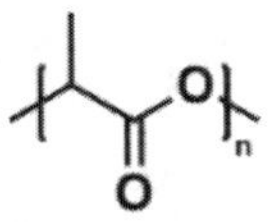

Figure 4.1. Molecular structure of polylactic acid.

Thermogravimetric analysis (TGA) is commonly used to determine the thermal degradation and mass variations of PLA and its (nano)composites. The weight loss below 150°C provides important details about the amount of water the specimen has absorbed. The decomposition temperature of PLA ranging between 300-400°C provides information on the rate of degradation [3]. In most cases, the presence of thermal conductive fillers enhances the thermal degradation rate due to their heat conduction. This process often facilitates char formation, thus protecting the underlying polymeric materials. The char residue can be analysed to determine the amount of filler present and how evenly it is dispersed throughout PLA. For example, a good filler dispersion produces char residue that is equal to the amount that was initially incorporated into the matrix.

The interaction between PLA and CBFs is an important factor that plays a crucial role in the resulting thermal properties [6]. Strong interaction between PLA and filler is reported to improve thermal stability of the composites [6]. This results from the filler acting as a barrier to protect polymer chains from thermal scission and thus improves thermal stability. It is worth mentioning that the information on the degradation stability can be used to provide knowledge on the processing parameters when using conventional melt-compounding techniques, especially processing temperature.

4.2. Thermal Properties

Currently, a variety of carbon (nano)fillers are being used as reinforcement of PLA, not only to improve thermal stability of PLA, but also other properties, such as electrical conductivity, mechanical performance and thermal conductivity [7]. The (nano)composites often exhibit different properties, depending on the filler-type, filler-to-filler interaction, polymer-filler interaction, and amount of filler [7, 8]. In addition, the processing methods are crucial in the developed composites. These methods can change all the previously mentioned factors, and thus the thermal behaviour of the (nano)composites. Table 4.1 summarizes the thermal behaviour of PLA/CBFs composites.

4.2.1. CNTs-Based Composites

CNTs are one of the most important carbon filler types that are widely used in preparation of PLA-CBFs composites. In spite of CNTs' presence to improve thermal stability of PLA, the interaction of CNTs with PLA plays an essential role in the extent of thermal stability [6, 9]. Strong interaction between the composite components, thus promoting filler distribution, is preferable to improve the overall thermal stability of the PLA. In this case, oxidation of CNTs to introduce oxygen functionalities has been the most commonly used method to improve interaction of CNTs with PLA. This cheaper modification involves the introduction of the CNTs into acid mixture, i.e., sulphuric and nitric acids, to introduce –COOH groups that are capable of forming hydrogen bonds with PLA (see Figure 4.2). For instance, thermal stability of PLA was found to be significantly improved in the presence of COOH-CNTs [9]. It was reported that the initial PLA decomposition temperature was ~261°C, with complete decomposition at ~361°C. In the presence of CNTs-COOH the initial temperature shifted to ~287°C. The strong interaction and good dispersion of COOH-CNTs in the PLA were essential for increasing thermal stability of the ensuing composite. The presence of CNTs with high thermal stability and thermal conductivity was the main reason for such improvement in thermal stability [10]. In general, CNTs act as heat barrier materials that protect PLA and thus delays PLA thermal decomposition [10]. Since the heating rate affects the thermal decomposition of polymer composites, it was found that the increase in heating rates resulted in shifting the decomposition temperature to higher values because of shorter time available for the sample to reach a given temperature [9]. The calculated activation energy of the composites justified that the presence of CNTs results in highly thermostable PLA composites. It is worth noting that the type of surface can influence the extent of thermal stability enhancement [10].

Table 4.1. Thermal properties of PLA composites

Formulation	Modification/filler content	Preparation method	Monitoring methods	T5% (°C)	ΔT5% (°C)	Td (°C)	ΔTd (°C)	Tg (°C)	Comments	Refs.
PLA	-	Solution-casting	TGA	280	-	337a	-		Due to good dispersion and strong interaction between composite components the thermal stability of PLA-g-CNTs-based composites was better than unmodified-based CNTs composites	[11]
PLA/CNTs	Unmodified 2wt%			309	+29	348a	+11			
PLA/ CNTs	PLA-grafted 2wt%			305	+25	355a	+18			
PLA		In-situ polymerization	DSC, TGA, POM	-	-	289		42.8	Crystallization rate and nucleation rate increased with increase in TRGO content due to chemical interaction between the composites' components. Thermal stability increased with increase in TRGO content	[12]
PLA/thermal reduced GO	0.01-2.00			-	-	281-299	-9-+10	47.4-51.6		
PLA	-	Solution-casting	TGA	-	-	-	-		No influence on the thermal stability of PLA regardless of the content for modified CNT-based composites, meanwhile reduction in thermal stability was observed for unmodified	[6]
PLA/CNTs	-/0.1-3%			-	-	-	-			
PLA/CNTs	COOH-CNTs (acid treated)/ 0.1-3%	Melt-pressing		-	-	-	-			

Formulation	Modification/filler content	Preparation method	Monitoring methods	T5% (°C)	ΔT5% (°C)	Td (°C)	ΔTd (°C)	Tg (°C)	Comments	Refs.
PLA/A-CNTs (silane treated)	Silane treated CNTs/0.1-3%			-	-	-	-		CNT-based composites	
PLA	-	Solution-casting	DSC, POM	-	-	-	-	66	Grafted-GO acted as nucleating sites for PLA which resulted in an increase in crystallinity due to strong interaction and better dispersion of the particles within the matrix	[13]
PLA/PLA-g-GO	PLLA-grafted GO/0.3-5%			-	-	-	-	62-63		
PLA	-	Solution-casting	POM, TGA, DSC	-	-	-	-	57	0.4wt% GO-ODA loading was sufficient to elevate crystallinity and thermal stability of PLA	[14]
PLA/ octadecylamine (ODA)-GO	Akyl (Octadecyl amine)/0.4%							58		
PLA	-	Melt-mixing	TGA, DSC	-	-	397	-	61.7	Between 8-12% CB loadings, good dispersion was achieved which improved overall properties of the composites	[15]
PLA/CB	4-20	Compression-moulding		-	-	401-413	+4-+16	61.9-63.4		
PLA	-	Melt-mixing	TGA	332	-	355	-	-	T5% reduced with increase in EG content and Td slightly increase with 5%EG loading	[16]
PLA/EG	5-10	Compression-moulding		306-330	-26-(-2)	342-358	-13-+3	-		

Table 4.1. (Continued).

Formulation	Modification/filler content	Preparation method	Monitoring methods	T5% (°C)	ΔT5% (°C)	Td (°C)	ΔTd (°C)	Tg (°C)	Comments	Refs.
PLA	-	In situ polymerization	DSC, TGA, POM	233		-	-	48.1	Regardless of the modification technique thermal properties were improved with the addition of hybrid filler improving the overall thermal properties of the composites when compared to individually reinforced composites	[17]
PLA/GO	Organically modified (-OH)/1wt%			241	+8	-	-	40.4		
PLA/(polyhedral oligomeric Silsesquioxane) POSS	Organically modified (-OH)/1wt%			245	+12	-	-	45.0		
PLA/GO-POSS	Organically modified (-OH)/1wt%			252	+19	-	-	59.5		
PLA/GO-g-POSS	1wt%			262	+29	-	-	58		
PLA pellets	-	Melt-compounding	TGA and DSC	373	-	377	-	-	Optimal content of EG to afford its dispersion is essential to achieve anticipated thermal stability	[3]
Processed PLA	-			372	-	378	-	62		
PLA/EG	-/8-12%			377-383	+5-+11	382-385	+4-+7	61-62		

Formulation	Modification/filler content	Preparation method	Monitoring methods	T5% (°C)	ΔT5% (°C)	Td (°C)	ΔTd (°C)	Tg (°C)	Comments	Refs.
PLA/PEG	-	Melt-compounding Melt-pressing	TGA and DSC	-	-	291	-	51.6	GNPs delays evaporation of volatile degradation products; hence improves the overall thermal stability	[18]
PLA/PEG/GNP	-/0.5-1wt%			-	-	344	+53	50.5-51.3		

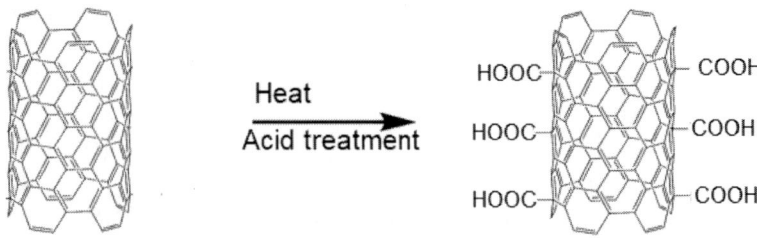

Figure 4.2. Functionalization reaction of CNTs.

Wu et al. [10] report that the thermal stability of CNTs/PLA is primarily dependent on the surface modification used. The authors compared the influence of pure CNTs, hydroxyl-decorated CNTs (CNTs-OH) and carboxyl-decorated CNTs (CNTs-COOH) on the thermal stability of PLA. Despite the composites exhibiting similar thermal stable residues corresponding to the initial amount of CNTs loaded, there was no improvement in temperature decomposition at 5% weight loss ($T_{5\%}$). This behaviour was attributed to two possible mechanisms: (i) CNTs hindering the crosslinking between PLA chains, and (ii) impurities from the CNTs synthesis acting as catalyst. The maximum degradation temperature (Td) for neat CNTs- and COOH-CNTs-based composites increased by 8-10°C. Besides poor dispersion of the neat CNTs, the high thermal conductivity and lack of reaction sites due to the absence of functional groups were responsible for such increase in thermal stability at a higher temperature. On the other hand, COOH-CNTs showed better dispersions, which facilitated the formation of percolation network to enhance overall thermal conductivity while acting as barrier to enhance thermal stability. The presence of hydroxyl groups produces Lewis or Bronsted acids sites on the surface of CNTs- promoted thermal depolymerisation of PLA. In addition, their poor dispersion reduced the barrier effect and thermal conductivity of CNTs and thus reduced the thermal stability of PLA. This demonstrates that the modification of CNTs has to be considered carefully in order to achieve the desired thermal stability. For example, the modification of CNTs using 3-isocyanatopropyl triethoxysilane (A-CNTs) improved their interfacial adhesion with PLA when compared to acid-treated CNTs and unmodified CNTs [6]. It was reported that unmodified CNTs reduced the thermal stability of PLA, whereas modified CNTs had no influence on the thermal stability of PLA, regardless of the filler content. This behaviour was attributed to strong interaction between LA and modified CNTs. The interfacial adhesion of the PLA/CBFs is often studied using differential scanning calorimetry (DSC). The glass transition temperature (T_g) obtained from DSC reflects the extent of interaction. The presence of rigid CNTs in PLA hinders the molecular chain mobility, which is reflected by shifting of T_g to higher temperatures when compared to neat PLA [19]. Therefore, the surface modification of CNTs often increases their interaction with PLA and thus increases the value of T_g. Yoon et al. [19] report on the grafting of PLA onto CNTs to improve their dispersibility and interaction with PLA. They report that the presence of CNTs increased the T_g of PLA. It was, however, noticed that modified CNTs further increased the T_g values due to good filler dispersion and strong interfacial adhesion between the composite components. These factors essentially hinder the chain mobility of PLA efficiently. In a similar manner, the crystallization behaviour of the resultant composite can be studied to provide insight into the interaction between PLA and filler. Using DSC, it was reported that cold-crystallization temperature was narrow and lower than that of neat PLA due to the nucleating efficiency of CNTs during crystallization [19]. The crystallization temperature increased with an increase in CNT-g-PLA, indicating that the crystallization rates increased due to the nucleating effect of CNTs resulting from strong interaction and good dispersibility. The melting temperature of PLA was not affected, but melting enthalpy increased in the presence of CNTs, though without a noticeable change in the CNTs content. Yet another important factor in the thermal stability of CBFs-reinforced PLA is the preparation method. The preparation method contributes to the dispersion and interfacial adhesion between CBFs and PLA, and thus their thermal stability (Table 4.1) [11].

A comparison between the effect of PLA-g-CNTs and unmodified CNTs incorporated into PLA using solution mixing on thermal properties was conducted by Kim et al. [11]. It reported that the presence of CNTs improved the thermal stability of the resulting composites when compared to neat PLA. The presence of PLA-g-CNTs, however, performed better than the unmodified CNTs. The calculated activation energies using the Kissinger and Flyn-Wall-Ozawa methods also confirmed the high thermostability of the composites. The activation energy for PLA-g-CNTs-based composites was higher than those of unmodified CNTs-based composites, which confirms that they are thermally more stable. From the estimates using the Kissinger method, the activation energy increased from 144 KJ mol^{-1} to 151 KJ mol^{-1}, whereas for Ozawa's methods it increased from 151 KJ mol^{-1} to 160 KJ mol^{-1}. This behaviour was ascribed to surface-grafted PLA improving CNTs dispersion and strong interfacial adhesion between CNTs and PLA. Elsewhere, an in-situ polymerization method was used to prepare PLA/CNTs composites [20]. The authors used traditional acid treatment to purify CNTs. Subsequently, the purified CNTs were treated with SOCl₂ in order to afford grafting of *L*-lactide using ring-opening polymerization (ROP). It was reported that the surface modification of CNTs improved their dispersion and interaction with PLA [20]. Thermal analysis methods were utilized to evaluate the extent of interaction between CNTs and PLA, which is important for overall performance of the resulting composite materials. In this case, the melting temperature of PLA decreased with an increase in PLA-g-CNTs because of strong interaction between PLA and CNT-g-PLA. [20] This was confirmed by calculating interaction energy density, B, using the following equation:

$$T_m^0 - T_{mix}^0 = -\frac{BV_{iu}}{\Delta H_{iu}} T_m^0 (1 - \emptyset_i)^2 \qquad 4.1$$

where $T^o{}_m$ and $T^o{}_{mix}$ are equilibrium melting temperatures for PLA and the mixture, respectively. $\Delta H_{iu}/V_{iu}$ is the heat of fusion of PLA per unit volume, while $\emptyset$ represents volume fraction. The value of B was found to be -96.6 cal cm^{-3}. The negative value of B confirms that there was favourable interaction between PLA and PLA-g-CNTs.

The presence of CNT and its content play a major role in the crystallization behaviour of PLA [21]. There is optimal content at which the crystallization can be enhanced. When the content of CNTs is increased beyond this concentration, the hindrance to retard crystallization is often observed. For instance, it was reported that the cold crystallization of PLA was enhanced, with an increase in CNT content up to 2% loadings [21]. The slight enhancement in cold crystallization was achieved when CNTs content was increased beyond 2%, indicating that this loading serves as percolation concentration. Beyond such concentration the fillers are no longer acting as nucleating site to facilitate the cold crystallization of PLA. The filler forms a network structure within the system corresponding to high melt viscosity, and thus hinders the crystallization of the host matrix.

The thermal stability of unmodified CNTs and modified CNTs-based 3D-printed components was investigated by Bortoli et al. [22]. The authors used oxidative conditions and found that PLA degraded at temperatures between 300°C and 400°C. Using decomposition temperature at initial weight loss of 5% ($T_{5\%}$) as parameter, it was found that the modified CNTs improved $T_{5\%}$ from 340°C for neat PLA, and unmodified CNT/PLA composites to 348°C. These results were associated with the good dispersion of modified CNTs and their strong interaction with PLA. This means that good dispersion and strong interfacial adhesion are crucial for the filler to physically act as shield to protect the underlying materials while delaying the transportation of the decomposition products, and thus enhancing thermal stability.

4.2.2. Graphene-Based PLA Composites

The thermal properties of graphene-based composites were reported to depend on the interaction between graphene and PLA as well as the dispersion of graphene nanosheets [13, 14, 16]. Moreover, the dispersion of graphene-based fillers was found to be directly dependent on the surface modification of graphene nanofillers and their content. $T_{5\%}$ of PLA decreased with an increase in expanded graphite; however, 5%EG loading increased T_d by 3°C. The latter was attributed to the barrier effect of the exfoliated graphene sheets against the volatile pyrolyzed products. The reduction of T_d with the further addition of EG beyond 5% was ascribed to three mechanisms: (i) in situ exfoliation of EG creating voids within the composite, (ii) the remaining acid from exfoliation of EG catalysing PLA degradation, and (iii) high thermal conductivity increasing heat conduction that promotes degradation of the composites. On the contrary, Murariu et al. [3] report that the incorporation of expanded graphite into PLA through melt-compounding resulted in highly thermally stable composites. It is recognized that expanded graphite is thermally stable from room temperature to 600°C under air, whereas PLA decomposition temperature is ~300°C. Thus, the presence of expanded graphite improved the thermal stability of PLA. $T_{5\%}$, $T_{50\%}$ and T_d increased with EG content due to the shielding effect resulting from the flake-like filler. The graphene sheets of EG serve as tortuous diffusion pathway of the volatile decomposition products. 8% EG loading was found to be optimal content to achieve maximum thermal stability, which with further increase leads to minor/no significant changes on the thermal stability. The effect of the dispersion of EG on the thermal stability was carried out by varying processing speed rates, i.e., 50, 100 and 150 rpm [23]. At 50 rpm for 5 minutes the composites displayed similar thermal degradation behaviours than the neat PLA. Thermal stability increases with residence time and speed rate, with the optimal processing extrusion parameter being 100 rpm and 10 minutes. This demonstrates that thermal stability is primarily dependent on the dispersion of EG.

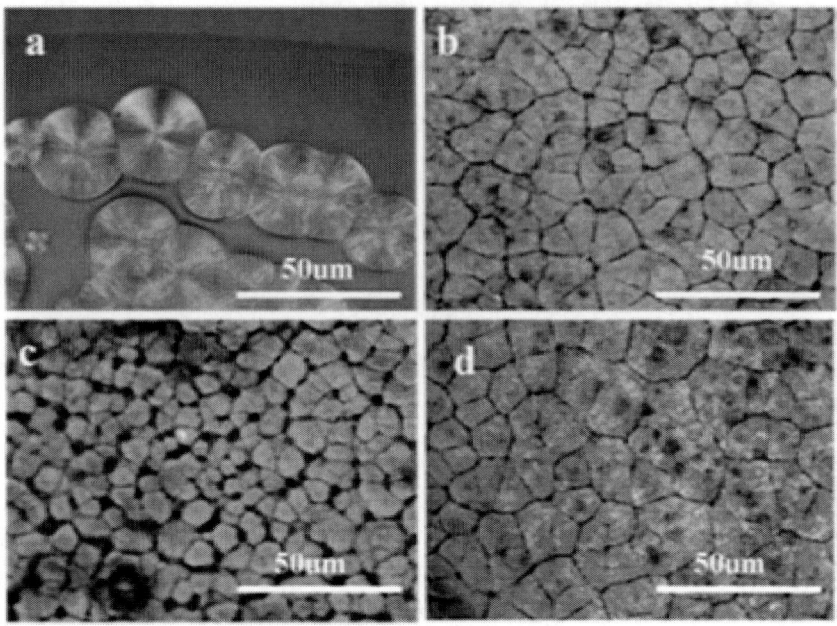

Figure 4.3. Polarized microscope (POM) images of isothermally crystallized PLA composites at 130°C: (a) PLA, (b) 1%PLA-g-GO/PLA, (c) 3%PLA-g-GO/PLA, and 5%PLA-g-GO/PLA. Reproduced from Wang et al. [13]. Open Access.

Zhang et al. [14] studied the influence of graphene oxide (GO) functionalization with long chain alkylamine (i.e., octadecylamine) (ODA) on the thermal behaviour of PLA. PLA and PLA/GO-ODA composites displayed single degradation steps between 300 and 400°C. The decomposition temperature at 5% weight loss for composites was ~14°C higher than that of neat PLA. This demonstrates that the presence of GO-ODA improves thermal stability of PLA due to the barrier effect of GO-ODA that delays the diffusion and escape of volatile degradation products. In addition, it promotes the formation of char. The strong interfacial adhesion and increase in crystallinity were also the contributors of high thermal stability of the composites. It was also found that the presence of functionalized GO resulted in a faster degradation rate due to the high thermal conductivity of reduced GO formed through the removal of oxygen functionalities at temperatures above 325°C. Wang et al. [13] report that strong interaction between PLA and graphene oxide (GO) was obtained after modification of GO with L-lactide. The melting temperature and crystallization temperatures were not significantly affected. The authors found that 3% loading of PLA-g-GO was sufficient to improve the crystallinity of PLA significantly, and beyond that a slight reduction was recorded. This was attributed to the agglomeration of fillers to form clusters that hinder the crystallization process of PLA. DSC curves showed two melting peaks in the presence of PLA-g-GO, indicating that there were two sizes of crystallites. The latter results from the incomplete crystallization of PLA due to fast crystallization of PLA in the presence of the GO. POM results indicated that PLA spherulites grew up to ~25 μm with clear boundaries, whereas a large number of smaller spherulites were formed in the presence of PLA-g-GO (Figure 4.3). This confirms the nucleating efficiency of GO particles. The number of spherulites decreased with an increase in GO loadings due to the agglomeration of the fillers.

The organic modification of GO nanosheets (OH-GO) resulted in the introduction of –OH groups that were exploited to graft L-lactide onto GO (PLA-g-GO) via in situ ROP [17]. The GO was also grafted onto polyhedral oligomeric silsesquioxane (POSS) nanofillers to afford a synergistic system for improving the thermal properties of the resulting composites. The authors introduced the fillers as well as the GO-g-POSS and GO/POSS individually into PLA. It was reported that Tg for PLA/GO-OH, and PLA/POSS was lower that of neat PLA, whereas the PLA/GO-g-POSS and PLA/GO/POSS recorded a ~10°C increment. The former was ascribed to the presence of lower molecular weight PLA chains. Melting temperatures of GO-OH and POSS-based composites were also found to be less than those of neat PLA. Meanwhile, GO-g-POSS and GO/POSS-based composites exhibited higher melting temperatures. All the composites' samples had higher crystallization temperatures with a narrow temperature range when compared to neat PLA, indicating that the nanofillers act as heterogeneous nucleating sites for PLA crystallization, which are more effective than the homogeneous nucleation process for neat PLA. This was confirmed by POM images for PLA samples with regular and round spherulites with a diameter of 50 μm, which correspond to the homogeneous crystallization process. In the case of composites, the spherulites were irregular and randomly spread out, indicating that fillers serve as nucleating sites for heterogeneous crystallization process. This confirms that the crystallization process was a bit shorter for all composites when compared to neat PLA. Higher crystallinity was attained for GO-g-POSS and GO/POSS-based composites, compared to neat PLA and PLA reinforced individually with GO and POSS as fillers. The glass transition temperature values of PLA were found to increase to higher temperatures in the presence of thermally reduced GO (TRGO) [12]. However, melting temperature values increased with an increase in TRGO content. The thermal stability of PLA was found to be dependent on the concentration of thermally reduced graphene oxide (TRGO). It was found that T_d increased from ~261 to 276°C with the addition of 1.5% TRGO. The initial thermal degradation of (T_i) of PLA was ~173°C, which increased to 190°C, 200°C, and 211°C for 0.01%, 1.0% and 2.0% TRGO loadings. The calculated time for a complete thermal decomposition was noticed to decrease with an increase in TRGO content because of the high thermal conductivity of graphene. This indicates that the incorporation of graphene can improve the thermal conductivity of the PLA.

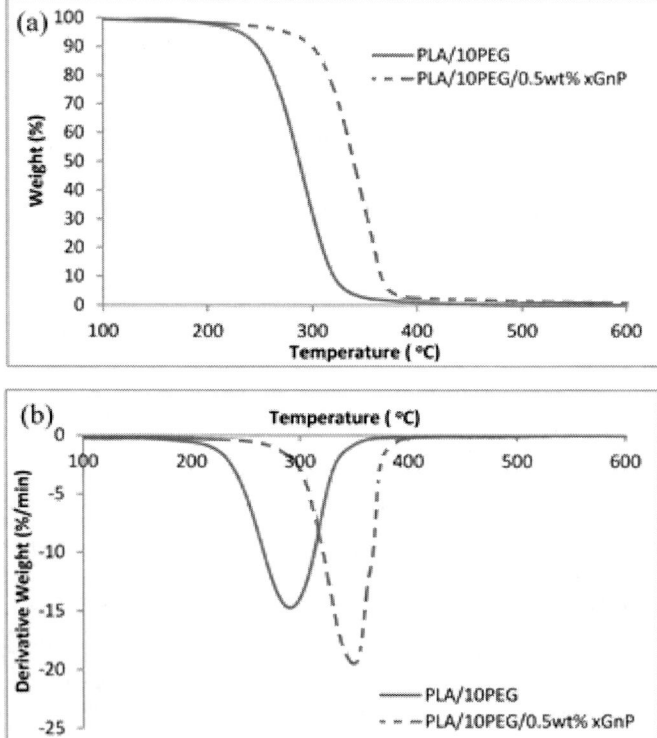

Figure 4.4. TGA and DTG curves of PLA/PEG and PLA/PEG0.5%GNPs. Reproduced from Chieng et al.[18]. Open Access.

Graphene nanoplatelets (GNPs) were introduced to polyethylene glycol (PEG) plasticized PLA to improve the overall thermal stability, as shown in Figure 4.4 [18]. T_d and the initial decomposition temperature increased from 291.0°C to 344.0°C, and from 194.5°C to 250.4°C with the addition of 0.3wt%, indicating that GNPs enhanced the thermal stability of PLA. This phenomenon was also attributed to the high thermal stability of GNPs and their capability for acting as a barrier effect responsible for delaying the evaporation of the volatile degraded products. In addition, the layered structure of graphene nanosheets from exfoliated GNPs serves as a tortuous path that hinders the diffusion of volatile degradation products and thus improves the thermal stability of PLA.

4.2.3. CB-Based Composites

Carbon black is one of the cheapest conductive enhancers in polymeric materials. Very few reports are based on the inclusion of CB into PLA for various applications. Recent study by Guo et al. [15] on the thermal properties of CB-reinforced PLA indicates that the content of CB plays a major role in the resulting properties. The glass transition was almost the same for all composites, regardless of the filler content. However, the initial degradation temperature and T_d was found to increase with CB content within 8-12% loadings. Initial degradation temperatures increased by 3-19°C, while Td increased by 4-16°C. This was attributed to better dispersion of the filler with a further increase beyond 12%, leading to filler agglomeration. The fillers agglomeration resulted in reduction in thermal stability of the resultant composites. In the same manner, da Silva et al. [24] report that there is optimal content to achieve maximum thermal stability of the resultant composite. They emphasize that 5–10%CB loading was adequate to increase PLA's thermal stability. The addition of 10%CB raised the PLA's thermal degradation temperature by 32°C. For all prepared composites Tg was constant indicating that there was limited interaction between PLA and CB particles. Blending PLA with polybutylene succinate (PBS) to modify the elasticity of PLA in the presence of CB as compatibilizer was reported in literature [25]. Despite the presence of CB slightly increasing Tg from ~65°C to 67°C, the presence of PBS reduced it to ~64.5°C. The incorporation of CB into the blend led to T_g that belongs to each blend's component to shift towards one another. This indicated that emulsification occurred in the presence of CB, resulting in thermodynamic interface stabilization.

4.3. Hybridization

Carbon nanofillers-reinforced composites received a great deal of interest due to their unique features, lightweightedness, ease of processing, and high strength [26]. The high-aspect ratio and specific area also contribute to these properties. The incorporation of the secondary filler into carbon nanofiller-reinforced composites have shown remarkable increments on the thermal, mechanical and electrical properties when compared to the neat polymers [26]. The resulting nanocomposites are suitable for high-end performance applications due to the resultant multifunctionality from the hybrid fillers' intrinsic properties. The combination of GNPs and carbon fibres as reinforcement of PLA was found to affect the residual mass of PLA [27]. The residue of PLA was found to be 47% at ~399°C, whereas the composite materials exhibited 8.8% residue at the same temperature. This was attributed to the reinforcement of carbon fibres and GNPs. The ratio of the fillers in hybrid composites plays a critical role on thermal properties [28]. There is optimal content of the filler at which the thermal stability of the materials can be improved drastically. For instance, it was reported that the inclusion of 0.1% GNPs into PLA/nanohydroxyapatite (NHA) significantly improved thermal stability; however, further loading led to a drastic reduction in thermal stability [28]. The thermal stability of the PLA reinforced with GO-OH and POSS increased by ~10°C, whereas it increased to ~19°C and ~31°C for PLA/GO/POSS and PLA/GO-g-POSS composites, respectively [17]. This demonstrates that there is synergy between GO and POSS in improving the thermal stability of PLA.

4.4. Conclusion

In general, the addition of CBFs enhances the PLA's thermal stability since these fillers are thermally more stable than PLA. The thermal behaviour of the resulting (nano)composites is significantly influenced by the interaction between the filler and PLA. There is optimal filler content to afford good interfacial adhesion and thus results in better thermal stability. Moreover, the dispersion of the fillers is directly dependent on the content of the filler, preparation method and surface modification of CBFs. The thermal stability and crystallization of polylactic acid are both improved by the homogeneous dispersion of the carbon nanofillers. The most often utilized CBFs modification techniques to improve dispersion and interfacial adhesion between the composite's components are acid purification and PLA-grafting onto CBFs. Besides the fact that the hybridization is the best method to enhance thermal stability of PLA, very few studies are based on the thermal properties of hybrid filler-reinforced PLA. One of these fillers acts as a linkage between PLA and other fillers, thus enhancing the composites' overall thermal properties.

References

[1] Mokhena TC, Sefadi JS, Sadiku ER, John MJ, Mochane MJ, Mtibe A. Thermoplastic Processing of PLA/Cellulose Nanomaterials Composites. *Polymers* 2018;10.

[2] Mokhena TC, Mochane MJ, Sadiku ER, Agboola O, John MJ. Opportunities for PLA and Its Blends in Various Applications. In: Gnanasekaran D, editor. Green Biopolymers and their Nanocomposites. Singapore: Springer Singapore; 2019. p. 55-81.

[3] Murariu M, Dechief AL, Bonnaud L, Paint Y, Gallos A, Fontaine G, et al. The production and properties of polylactide composites filled with expanded graphite. *Polymer Degradation and Stability* 2010; 95:889-900.

[4] Lim LT, Auras R, Rubino M. Processing technologies for poly(lactic acid). *Progress in Polymer Science* 2008; 33:820-52.

[5] Mochane MJ, Mokhena TC, Sadiku ER, Ray SS, Mofokeng TG. Green Polymer Composites Based on Polylactic Acid (PLA) and Fibers. In: Gnanasekaran D, editor. *Green Biopolymers and their Nanocomposites.* Singapore: Springer Singapore; 2019. p. 29-54.

[6] Chiu W-M, Kuo H-Y, Tsai P-A, Wu J-H. Preparation and Properties of Poly (Lactic Acid) Nanocomposites Filled with Functionalized Single-Walled Carbon Nanotubes. *Journal of Polymers and the Environment* 2013; 21:350-8.

[7] Silva VAOP, Fernandes-Junior WS, Rocha DP, Stefano JS, Munoz RAA, Bonacin JA, et al. 3D-printed reduced graphene oxide/polylactic acid electrodes: A new prototyped platform for sensing and biosensing applications. *Biosensors and Bioelectronics* 2020; 170:112684.

[8] Huang X, Yin Z, Wu S, Qi X, He Q, Zhang Q, et al. Graphene-Based Materials: Synthesis, Characterization, Properties, and Applications. *Small* 2011; 7:1876-902.
[9] Chrissafis K, Paraskevopoulos KM, Jannakoudakis A, Beslikas T, Bikiaris D. Oxidized multiwalled carbon nanotubes as effective reinforcement and thermal stability agents of poly(lactic acid) ligaments. *Journal of Applied Polymer Science* 2010; 118:2712-21.
[10] Wu D, Wu L, Zhang M, Zhao Y. Viscoelasticity and thermal stability of polylactide composites with various functionalized carbon nanotubes. Polymer Degradation and Stability 2008; 93:1577-84.
[11] Kim H-S, Hyun Park B, Yoon J-S, Jin H-J. Thermal and electrical properties of poly(l-lactide)-graft-multiwalled carbon nanotube composites. *European Polymer Journal* 2007; 43:1729-35.
[12] Yang J-H, Lin S-H, Lee Y-D. Preparation and characterization of poly(l-lactide)–graphene composites using the in situ ring-opening polymerization of PLLA with graphene as the initiator. *Journal of Materials Chemistry* 2012; 22:10805-15.
[13] Wang L-N, Guo Wang P-Y, Wei J-C. Graphene Oxide-Graft-Poly(l-lactide)/Poly(l-lactide) Nanocomposites: Mechanical and Thermal Properties. *Polymers* 2017.
[14] Zhang L, Li Y, Wang H, Qiao Y, Chen J, Cao S. Strong and ductile poly(lactic acid) nanocomposite films reinforced with alkylated graphene nanosheets. *Chemical Engineering Journal* 2015; 264:538-46.
[15] Guo J, Tsou C-H, Yu Y, Wu C-S, Zhang X, Chen Z, et al. Conductivity and mechanical properties of carbon black-reinforced poly(lactic acid) (PLA/CB) composites. *Iranian Polymer Journal* 2021; 30:1251-62.
[16] Xue B, Ye J, Zhang J. Highly conductive Poly(L-lactic acid) composites obtained via in situ expansion of graphite. *Journal of Polymer Research* 2015; 22:112.
[17] Pramoda KP, Koh CB, Hazrat H, He CB. Performance enhancement of polylactide by nanoblending with POSS and graphene oxide. *Polymer Composites* 2014; 35:118-26.
[18] Chieng BW, Ibrahim NA, Yunus WM, Hussein MZ. Poly(lactic acid)/Poly(ethylene glycol) Polymer Nanocomposites: Effects of Graphene Nanoplatelets. *Polymers* 2014. p. 93-104.
[19] Yoon JT, Jeong YG, Lee SC, Min BG. Influences of poly(lactic acid)-grafted carbon nanotube on thermal, mechanical, and electrical properties of poly(lactic acid). *Polymers for Advanced Technologies* 2009; 20:631-8.
[20] Chen G-X, Kim H-S, Park BH, Yoon J-S. Synthesis of Poly(L-lactide)-Functionalized Multiwalled Carbon Nanotubes by Ring-Opening Polymerization. *Macromolecular Chemistry and Physics* 2007; 208:389-98.
[21] Xu Z, Niu Y, Yang L, Xie W, Li H, Gan Z, et al. Morphology, rheology and crystallization behavior of polylactide composites prepared through addition of five-armed star polylactide grafted multiwalled carbon nanotubes. *Polymer 2010*; 51:730-7.

[22] Bortoli LSD, Farias Rd, Mezalira DZ, Schabbach LM, Fredel MC. Functionalized carbon nanotubes for 3D-printed PLA-nanocomposites: Effects on thermal and mechanical properties. *Materials Today Communications* 2022; 31:103402.
[23] Hassouna F, Laachachi A, Chapron D, El Mouedden Y, Toniazzo V, Ruch D. Development of new approach based on Raman spectroscopy to study the dispersion of expanded graphite in poly(lactide). *Polymer Degradation and Stability* 2011; 96:2040-7.
[24] Silva TFd, Menezes F, Montagna LS, Lemes AP, Passador FR. Preparation and characterization of antistatic packaging for electronic components based on poly(lactic acid)/carbon black composites. *Journal of Applied Polymer Science* 2019; 136:47273.
[25] Wang X, Zhuang Y, Dong L. Study of carbon black-filled poly(butylene succinate)/polylactide blend. *Journal of Applied Polymer Science* 2012; 126:1876-84.
[26] Jyoti J, Singh BP. A review on 3D graphene–carbon nanotube hybrid polymer nanocomposites. *Journal of Materials Science* 2021; 56:17411-56.
[27] Mohammed Basheer EP, Marimuthu K. Carbon fibre-graphene composite polylactic acid (PLA) material for COVID shield frame. *Materialwissenschaft und Werkstofftechnik* 2022; 53:119-27.
[28] Michael FM, Khalid M, Chantara Thevy R, Raju G, Shahabuddin S, Walvekar R, et al. Graphene/Nanohydroxyapatite hybrid reinforced polylactic acid nanocomposite for load-bearing applications. *Polymer-Plastics Technology and Materials* 2022; 61:803-15.

Chapter 5

Flammability Properties of PLA/Carbon Based Filler(s) Composites

Abstract

Poly(lactic acid) (PLA) is a biodegradable polymer that is produced from renewable resources and it is the polymer of interest in many engineering applications such as electrical devices and automotive applications. However, for PLA to be used in such applications, it needs to be tailored in order to meet the commercial standards. One of the main disadvantages of PLA is the high flammability of the biopolymer. In order to improve the flame resistance properties of the PLA, various flame-retardant fillers are incorporated into the PLA matrix. The carbon-based fillers are preferred flame-retardant fillers due to their cost and effective flame retardancy. This chapter discusses the flammability properties of PLA/carbon-based fillers. Factors such as the type of carbon-based filler, synergistic effect and the medications of the flame-retardant fillers are discussed.

Keywords: carbon based fillers, flame retardancy, protective char, bioplastics, halogen free-flame retardant fillers

5.1. Introduction

There is a need for the development of new greener products in order to reduce the dependence on the fossil fuel [1]. The utilization of the bioplastics is offering a solution to the environmental threat that is posed by the non-biodegradable polymers. Poly(lactic acid) (PLA) is one of the most popular biopolymers in the bioplastic family due to its numerous advantages such as biodegradable, easily processed, and biocompatible [2, 3]. However, there are inherent disadvantages that are associated with the PLA such as highly flammable [4] and low thermomechanical stability [5]. In order to obtain a durable structural material with improved flammability resistance properties, various nanomaterials are incorporated into the PLA matrix (Figure 5.1). In the past, halogenated flame-retardant fillers were utilized in order to enhance the flame retardancy of the polymers [6-8].

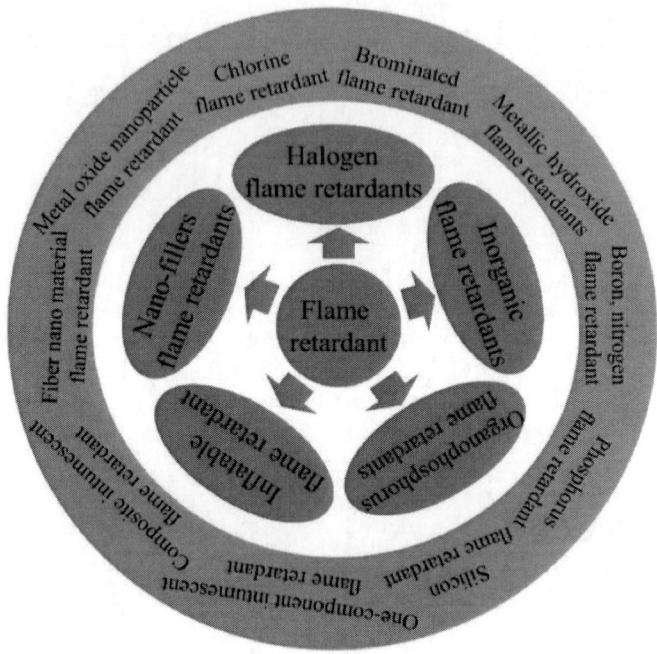

Figure 5.1. Various flame-retardant fillers [9].

Halogenated flame-retardant fillers are known to be the most effective flame-retardant fillers. However, currently the use of halogenated flame causes huge harm to the environment, which has limited the utilization of the halogenated flame-retardant fillers for advanced applications. For example, halogenated flame-retardant fillers release toxic smoke and gaseous hydrogen halide during the burning process [9]. This has prompted researchers to find alternative halogen-free flame-retardant materials. Carbon-based flame retardants have emerged as clear favourites to enhance the flammability resistance of the polymer matrices due to their low cost and effective flame retardancy with low smoke generation and anti-dripping behaviour [10]. The current chapter discusses the effect of various carbon-based fillers as flame-retardant fillers for PLA matrix. Furthermore, the synergistic effect of carbon-based fillers with other flame-retardant fillers is also discussed in depth. Figures 5.2, 5.3 and 5.4 present the bibliometric data extracted from the Web of Sciences (WoS) core collection databases using the keywords "Flame retardancy of PLA". Figure 5.2 shows a consistent level of publications from the years 2019 to 2021, followed by a sharp spike in research productivity in 2022. There is a steep decline in 2023, which can be attributed to the time the analysis was conducted, i.e., before the 2023 year-end.

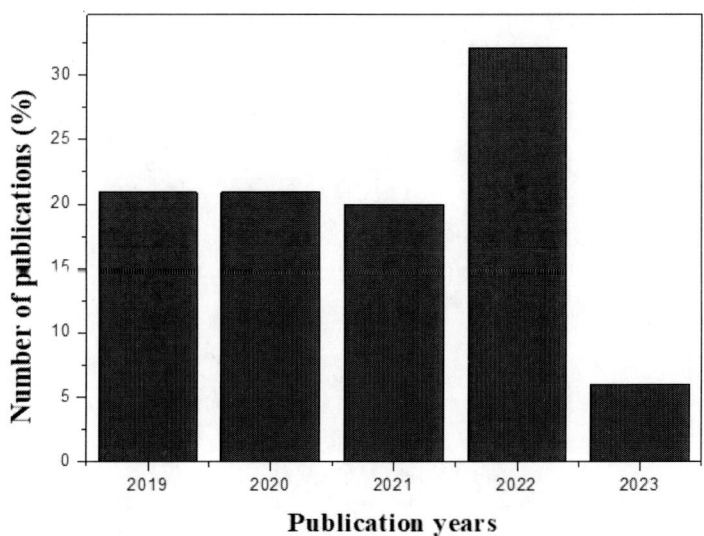

Figure 5.2. Flame retardancy of PLA publication trends in the last five years (16 April 2023).

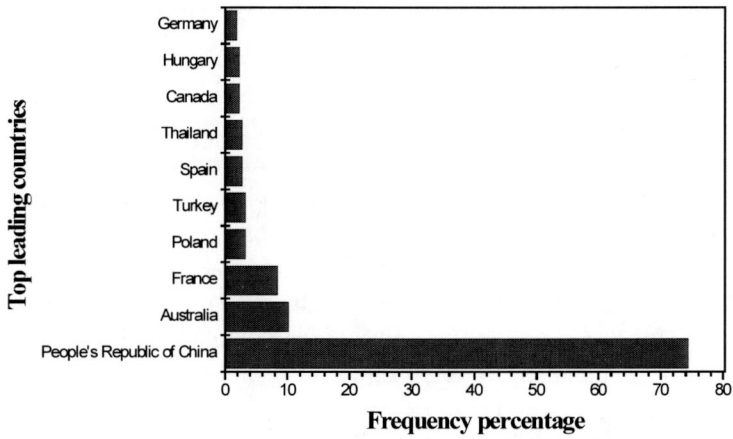

Figure 5.3. Top 10 leading countries in flame retardancy of PLA research.

Figure 5.3 shows that the People's Republic of China is the leading country in flame retardancy of PLA research (74.4%), followed by Australia (10.2%) and France (8.4%), respectively.

Figure 5.4 shows the visualization network of the keywords that have often co-occurred together in publications.

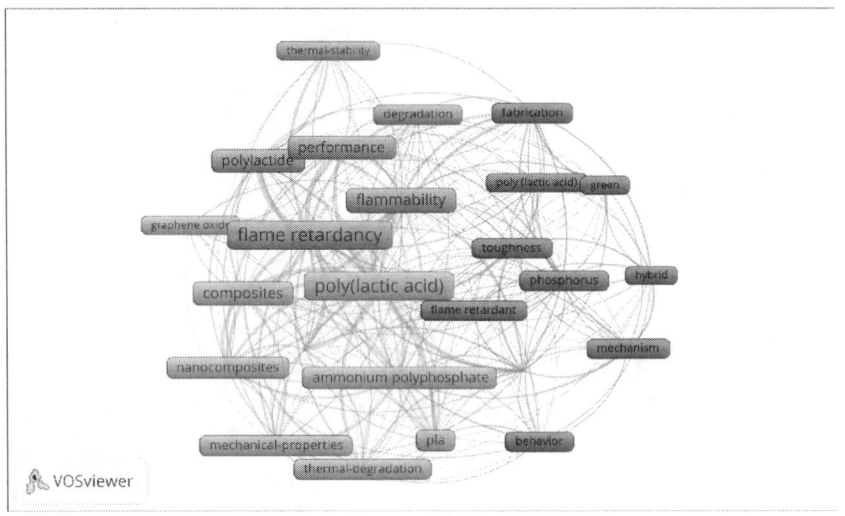

Figure 5.4. Bibliometric network visualization of the co-occurrence of keywords in PLA flame-retardancy research.

5.2. Flammability Properties: Graphite and Its Derivates/PLA Based Nanocomposites

The polymer nanocomposites incorporated with carbonaceous fillers has proven to enhance the fire resistance of the polymer matrices [11]. This type of flame-retardant fillers has the ability to form a protective layer during combustion, which inhibits the transfer of heat from the flame to the substrate and transfer of heat from the substrate into the flame. Thus, this mechanism has the tendency of decreasing the overall rate of flame feeding from the volatile products of the polymer pyrolysis and thermo-oxidation [12-15]. This mechanism is such that when PLA matrix undergoes a combustion, there would be a dripping of the polymer flaming which enhances the progression fire, whereas carbon-based fillers form a barrier which obstructs combustible gases and prevents further ignition of the polymer matrix. The flammability properties of composites were analysed by a cone calorimeter, limiting oxygen index (LOI) and UL-94. Figure 5.5 illustrates a schematic representation of the cone calorimeter techniques with the major component of the technique.

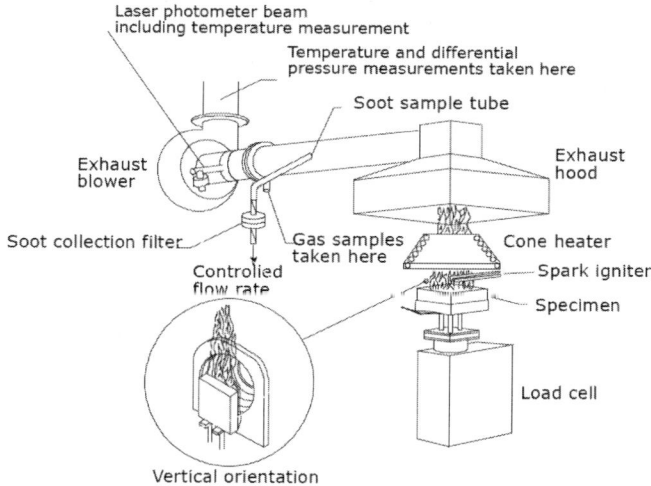

Figure 5.5. Schematic representation of the cone calorimeter indicating the major components [16].

Murariu et al. [11] report on the properties of the PLA reinforced with expanded graphite. One of the properties investigated by the authors was the flammability properties of PLA/graphite composites. The flammability properties of the PLA/expanded graphite with a composition of 3, 6% and 12% were investigated by the UL-94 and a cone calorimeter. According to the UL-94 results, neat PLA was observed to have dripped with no char, while the PLA/EG composites burned with little dripping (PLA-3% EG), and in some cases there was no dripping (PLA-6% EG and 12%EG). The results (UL-94 HB) were supported by the cone calorimeter observations, whereby the neat PLA revealed a high peak heat release rate (pHRR), the addition of EG at 6% reduced the pHRR by approximately 30%. This behaviour may be ascribed to the swelling of the graphite during burning and the formation of the porous foamed carbonaceous char, which acted as a barrier to protect the substrate against heat. Expandable graphite was utilized by Melillo and co-workers as a flame-retardant filler for the PLA the matrix [17].

Table 5.1. UL-94 test for all the investigated samples [17]

Sample	T_1	T_2	T_1+T_2	T_2+T_3	Flaming particles	Classification
PLA-EG0%	<10	***	<250	***	Yes	None
PLA-EG1%	<30	<30	<50	<60	Yes	V-2
PLA-EG3%	<30	<30	<250	<60	Yes	V-2
PLA-EG5%	<10	<10	<50	<30	Yes	V-2

T = time in seconds. *** not applicable.

The content of expandable graphite was incorporated into the PLA matrix within the range of 1 to 5% and the composites were prepared by extrusion. The flammability property of the PLA/expandable graphite was analysed by UL-94 (Figure 5.6) and the results are summarized in Table 5.1. The neat PLA did not pass the UL-94, since there was a dripping of the material. However, all the composites reached V-2 rating and various reasons were provided for such an observation. It was suggested that a lack of adhesion between the phases might have contributed to the rating of UL-94.

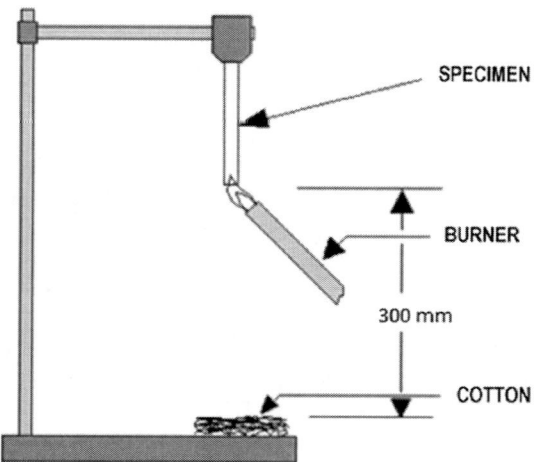

Figure 5.6. A typical setup for UL-94 [18].

The synergistic effect of graphite with other nanoparticles reinforced in PLA played a critical role in the flammability properties of the resultant composites. Fukushima et al. [19] report on the effect of both the expanded graphite and layered silicate clay on the fire-retardant properties of the PLA. The vertical (UL-94 V) and horizontal burning tests (UL-94 HB) were used for analysis flammability properties of the PLA and its composites. Unlike in the previous studies discussed above, the PLA in this study surprisingly has a V-2 rating. This is an indication that the PLA showed a slight fire retardance with a polymer residual of 64%. The addition of the nanofillers heavily modifies the burning process. The addition of clay into PLA (According to the UL-94 HB) matrix formed a char layer that is able to prevent the dripping of the composite. The burning rate (BR) of the composite was found to be higher than the neat PLA, and the classification thereof was termed the "limit" flammability classification. The binary composites based on EG melted without dripping with the formation of an intumescent char, with the char becoming more prominent with increasing in EG content, reaching a "limit" classification. The ternary composites (co-addition of the fillers) were also analysed by UL-94 HB. The ternary composites presented obvious anti-dripping properties as a result of the formation of a compact char accompanied by low BR when compared with neat PLA and binary compositions. The UL-94 V supported the UL-94 HB, whereby the incorporation of the clay and EG into the PLA showed a decrease in the BR when compared with the neat PLA. This is associated with the production of the carbonized-silicate char layer that is able to act as a barrier protecting the substrate against heat. The synergistic effect of expandable graphite and ammonium polyphosphate (APP)-reinforced polylactide (PLA) was reported by Zhu et al. [20]. The composites were fabricated by a Brabender mixer at the temperature of 180°C at a speed of 50 rpm for a duration of 8 min. PLA-based composites incorporated with 15wt% of APP/EG (1:3) were reported. The flammability of the composites was evaluated by UL-94 tests, LOI tests and cone calorimeter. Neat PLA was reported to have a LOI value of 22, whereas the LOI value of the PLA/APP was revealed to be 27.1. The highest LOI value (*viz* 41) amongst the single filler was obtained for the composite system consisting of PLA/EG with 15wt% of EG. This observation is a clear indication that the EG improves the LOI value of the PLA better than the APP. The synergy of 20wt% of the additives at an APP:EG ratio of 1:3 obtained the highest LOI value of 47.5 with a UL-94 V-0 rating. The cone calorimeter showed a PHHR value of 272 kW/m^2, while there was a reduction with the PLA-5 and PLA-9, with 208.4 and 167.9 kW/m^2, respectively. The reductions are associated with the formation of a higher-quality char layer. Such char layers are known for protecting the matrix from heat and oxygen, as a result reducing the PHHR values. A flame-retardant performance of a synergy between aluminium hypophosphite and expanded incorporated into the Bio-based polylactic acid were reported by Tang and co-workers [21]. A flame-retardant biodegradable PLA consisting of aluminium hypophosphite (AHP) and expanded graphite (EG) was fabricated by melt-compounding. The nanofillers were used in the composition range of 10-20wt% within the PLA matrix. As expected from the cone calorimeter, the PLA was found to burn faster and a peak appeared with a pHRR value of *ca.* 549 kW/m^2, while the addition of AHP with a composition of 10wt% and 20wt% decreased the pHRR values to 368 and 285 kW/m^2, respectively. The synergy of AHP and EG at 20wt% reduced the pHRR value by 52.3% when compared with neat PLA. One of the parameters for cone calorimeter is the total heat release (THR). The THR value for PLA was revealed as 62.3 KJ/m^2, while the addition of the AHP had a slight change in the THR values of the PLA composites. A synergy of fillers reduced the THR values when compared with neat PLA; for example, the THR of the PLA/15AHP/5EG composite was 49.8 KJ/m^2 when compared with 62.3 KJ/m^2 for neat PLA. Currently, there is a huge shift in terms of environmentally friendly flame-retardant fillers. Yang and co-workers [22] produced a novel flame-retardant filler by self-assembling melamine and phytic acid (PA) into a reduced graphene oxide (rGO) and the filler was introduced into the PLA matrix. The fabricated filler was defined as the PMrG. Unlike the previous studies, the composites in this study were fabricated by solution mixing utilizing dichloromethane. Firstly, the PMrG was mixed with dichloromethane and then the PLA powder was introduced into the mixture. The flame retardancy of the resultant composite was investigated by cone calorimeter tests (CCT), limiting oxygen index (LOI) and vertical burning test (UL-94). The neat PLA had an LOI value of 19% and it did not pass the UL-94 test, with no grade. The introduction of rGO into the PLA matrix resulted in an enhancement of LOI value of 22%, while the incorporation of 10wt% of PMrG revealed 25% of the LOI with V-0 rating, which proved that a synergy of rGO and PM to form PMrG was very effective in terms of enhancing the flame retardancy. According to the cone calorimeter results, a parameter such as HRR was observed to decrease with the addition of rGO and PMrG into the PLA matrix. For example, the HRR of PLA was recorded as 426.6 kW/m^2, while the PLA/rGO showed a 27% reduction in HRR with a value of 309.1 kW/m^2. A 10wt% of phytic acid-melamine (PM) revealed a higher pHRR value when compared with neat PLA. However, there was immense reduction in the THR of the PLA/PM when compared with neat PLA. The PLA/PMrG composite recorded the lowest HRR value of 276.1 kW/m^2, which confirmed that the PMrG exhibited the best flame retardancy at the ratio: rGO: melamine:PA 1:1:5. Another important parameter that is evaluated through the cone calorimeter technique is the smoke production rate (SPR) and total smoke rate (TSR). The release of smoke by burning materials plays a huge role in terms of death or casualties during burning. It was noted that the neat PLA had little smoke release when compared with rGO or PM. A higher smoke release in the presence of rGO and PM is associated with the incomplete combustion of the flame-retardant fillers. Interestingly, the PMrG/PLA composite resulted in a decrease in smoke release, which proved that their ratios at specific optimum ratio has the ability to reduce the smoke release. The difference in flammability of the investigated samples was attributed to their differences in char residues. The neat PLA formed a little or no char residues (Figure 5.7 a, e), which led to a weak barrier effect accompanied by high HRR peak. A better char was formed in the presence of rGO when compared with neat PLA (Figure 5.7b, f). However, the formed char was also associated with crack and cavities. The PM/PLA composites also presented a weak char layer, which is confirmed by high HRR value. However, the PMrG/PLA composite produced a dense and intact char layer, which can easily prevent the transfer of heat and supply of combustible gases during combustion, and as a result reducing parameters such HRR, THR and TSR values.

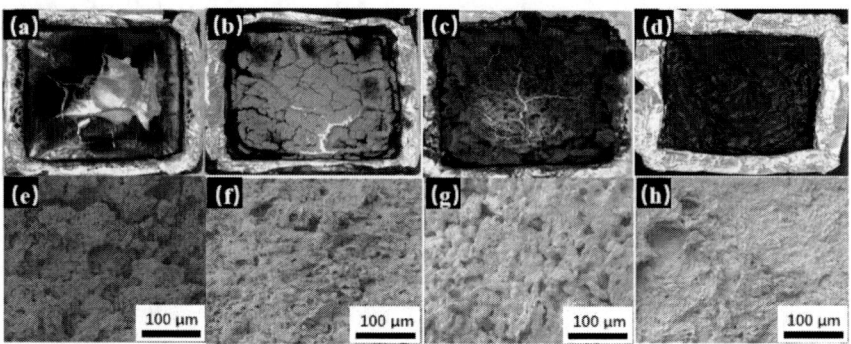

Figure 5.7. Digital images (char residues) of: (a) PLA, (b) rGO/PLA, (c) PM/PLA, and (d) PMrG/PLA and their SEM: (e) PLA, (f) rGO/PLA, (g) PM/PLA, and (h) PMrG-/PLA [22].

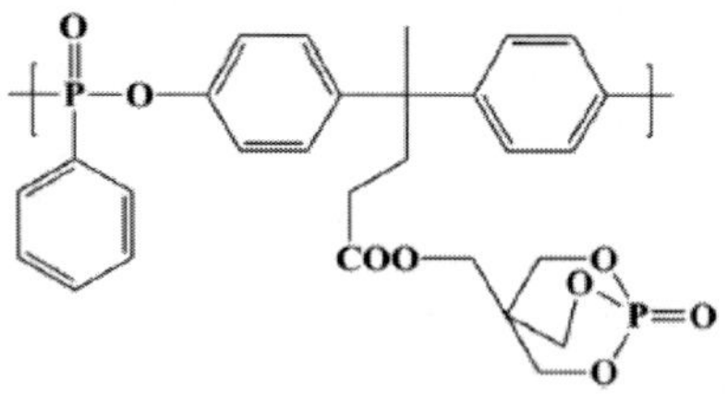

Figure 5.8. The chemical structure of bio-based polyphosphate [23].

Graphene oxide (GO), which is one of the derivatives of graphite, has also been used as a flame retardant for PLA matrix. Jiang et al. [23] report on the synergy of bio-based polyphosphate (BPPT) (Figure 5.8) and modified graphene oxide (GO) as flame-retardant fillers for poly(lactic acid) (PLA).

Modification of GO was done by grafting Polyethylenimine (PEI) into the GO in presence of 1-ethyl-3-(3-dimethylaminopropyl) carbodiimide hydrochloride (EDC.HCl).

The flame-retardant tests such as limiting oxygen index (LOI), vertical grade (UL94) and cone calorimeter have been used to test the flammability properties of polymers. The LOI value of neat PLA was reported to be 20 and as it has been extensively mentioned in this chapter, that PLA had no grade for UL-94. The incorporation of 3wt% BBPT in the matrix increased the LOI value of the composite to 33.6 and a UL-94 rating when compared with neat PLA. The BPPT flame-retardancy mechanism occurs through a gas phase, whereby there is a formation of phosphorus-containing compounds, which are well-known radical scavengers in the gaseous production in the process of thermal degradation of the flame-retardant filler (BPPT). The synergy of the M-GO and BPPT recorded the highest LOI value of 36.0 and a rating of V-0 from the UL-94. It was observed that the PLA/BPPT/M-GO composite showed no melt-dripping. The anti-dripping of the composites is associated with the high viscosity of the graphene nanosheets. The mechanism of graphene oxide takes place in the condensed phase, whereby the M-GO reduces the mass transfer and inhibits the dripping process. However, it became evident that the BPPT affects the flammability resistance of PLA through the gas phase. In this system, M-GO plays a huge role in reducing the decomposed the gas products that emanate from the BPPT, which may also affect the role of BPPT as a gas-phase type of flame-retardant filler. The results were further supported by the cone calorimeter, whereby the pHRR and THR values were recorded as 393 kW/m^2 and 67.1 MJ/m^2, respectively. The PLA/3 BPPT composites revealed a decrease in pHRR (i.e., 370 kW/m^2) and THR (*viz* 65.6 MJ/m^2), which is evidence that the BPPT enhances the flame resistance of the PLA. It was further observed that the addition and increase in total smoke release (TSR) reduced the TSR values from 216 (PLA/3 BPPT) to 112 m^2 m^{-2} (PLA/2.1BPPT/0.9 M-GO). GO is capable of acting as a good gas suppresser, forming a gas barrier delaying and trapping the diffusion of flammable gas by products. The behaviour is well supported by the SEM photos (Figure 5.9), which revealed a tactoid-layered structure, which has the ability to inhibit the movement of fuels produced by an easily flammable PLA matrix.

The char analysis of the investigated samples from cone calorimetry revealed thin chars with clear cracks (Figure 5.10).

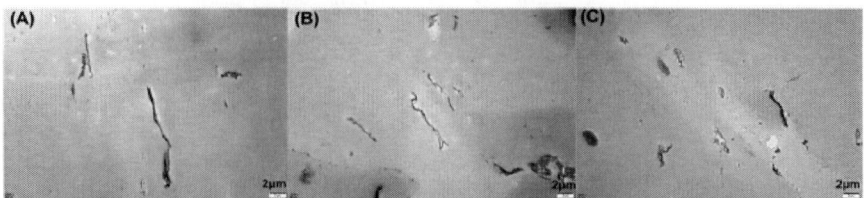

Figure 5.9. SEM photos of the samples: (A) PLA/2.7 BPPT/0.3-GO, (B) PLA/2.4 BPPT/0.6M-GO and PLA/2.1BPPT/0.9M-GO [23].

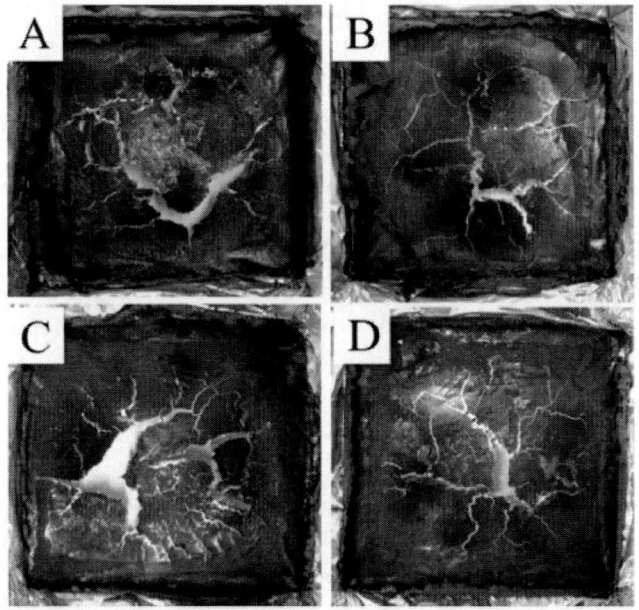

Figure 5.10. Digital photos of: the char residues for: A) PLA/3BPPT, B) PLA/2.7 BPPT 0.3 M-GO, C) PLA/2.4 BPPT/0.6 M-GO, and PLA/2. 1BPPT/0.9-GO [23].

The chars were further analysed by SEM (Figure 5.11), and it was observed that the PLA/3BPPT revealed a weak and loose char residual with a lot of holes. However, the presence of M-GO (0.3wt%) showed a continuous char layer with a lesser degree of weakened char layer. The incorporation of 0.6wt% of the M-GO resulted in a compact char, which is known to be a good protective char layer against the entrance of heat into the system and traps the volatiles materials out of the system as they might act as degradation catalysts. Beyond the 0.6wt% content of M-GO, i.e., 0.9wt%, there is a fabrication of the discontinuous char layer, with apparent holes, which resulted in a decrease in flame resistance of the system.

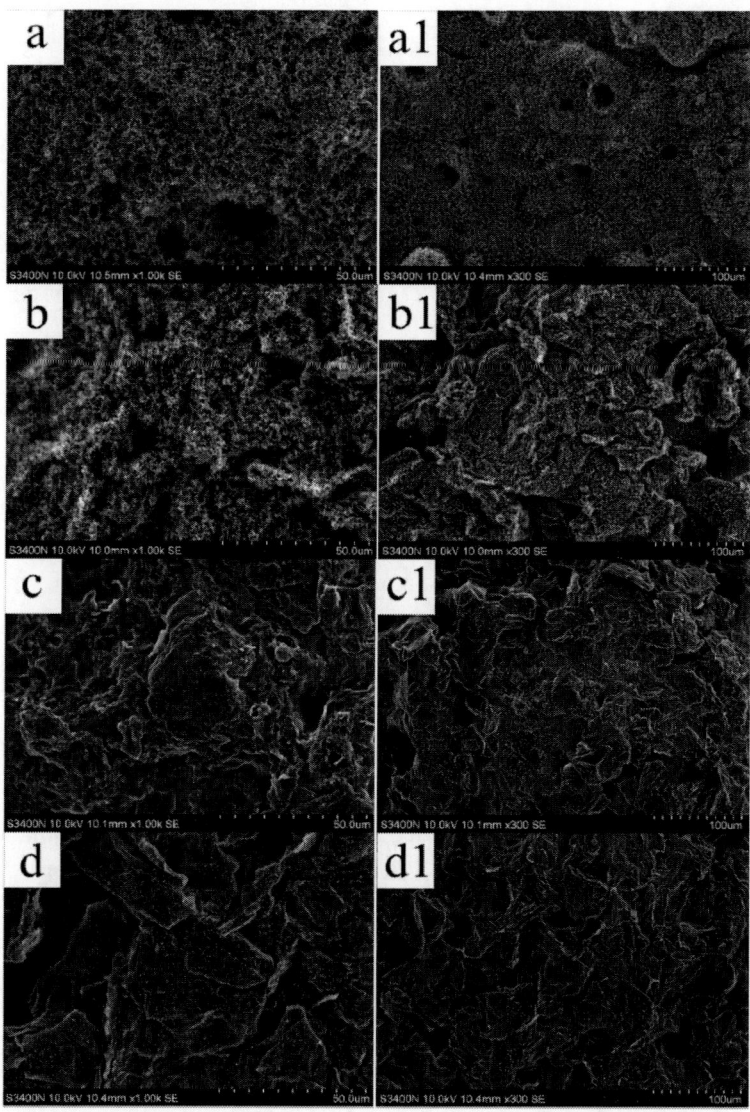

Figure 5.11. SEM of the char residues for: (a and a1) PLA/3BPPT, (b and b1) PLA/2.7 BPPT/0.3M-GO, (c and c1) PLA/2.4 BPPT/0.6 M-GO, and (d and d1) PLA/2.1BPPT/0.9 M-GO [23].

5.3. Flammability Properties: PLA/Carbon Nanotubes Composites

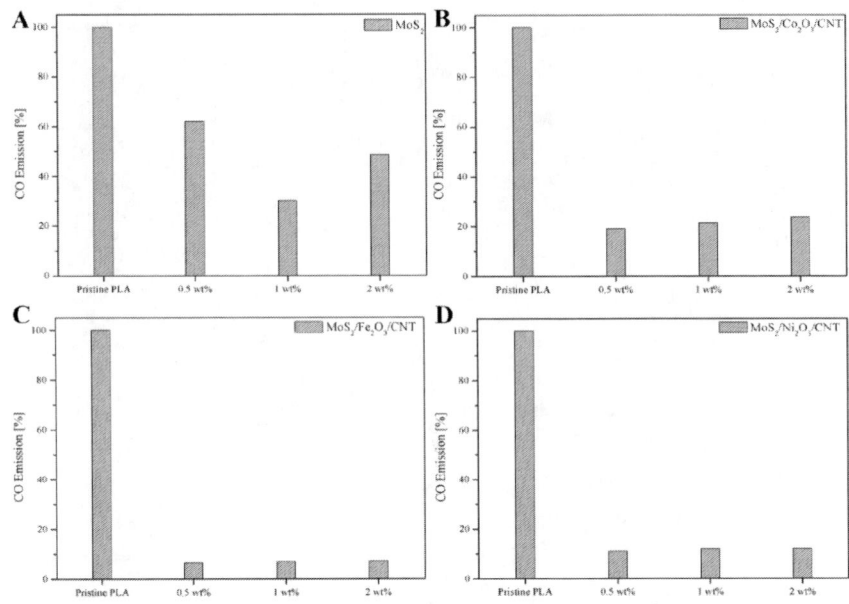

Figure 5.12. Carbon monoxide emission of PLA and its composites [24].

Carbon nanotubes (CNTs) have been utilized to enhance the flame retardancy of the polymer matrix. CNTs have been found to enhance the HRR (indicated by reduction in peaks) and mass loss rate (MLR) of the polymer composites. Factors such as the content of carbon nanotubes, chemical modification of CNTs and its synergy with other nanofillers have been found to affect the flammability properties of polymer/carbon nanotubes composites. Guo et al. [25] report on the flammability of the modified carbon nanotubes in the poly(lactic acid). The multi-walled carbon nanotubes (MWCNTs) were functionalized by phosphaphenanthrene compounds (9,10-dihydro-oxa-10-phosphaphenanthrene-10-oxide (DOPO) and vinyl triethoxy silane (VTES) and the modified MWCNTs termed the DVCNT. The flame-retardant PLA-based composites were fabricated by mixing CNT or DVCNT with a weight % of 1-4 by a rheometer 190°C for a duration of 6.5 min. According to the cone calorimeter, the incorporation of the CNTs and DVCNTs reduced the burning of the PLA to some extent, with the impact of DVCNT having more effect than the CNTs. The flame retardancy between the PLA/CNTs_4% and PLA/DVCNT_4% was compared and it was realised that the charring and barrier property of the PLA/DVCT_4% was more compact and stronger than the PLA/CNTs_4%. This was ascribed to the interaction between the pyrolysis of the DVCNT, which enhanced the charring and barrier effect of the char. This type of char is more effective in preventing heat from entering the system and also blocks the volatiles from leaving the system. The observation was further supported by the SEM images of the obtained chars, whereby the PLA/CNTs_4% image showed a cracked and discontinuous char, while the image for PLA/DVCNT_4% revealed a compact and complete char that is competent enough to stand against heat and improved the flame retardancy. It has been noted elsewhere [26] in the literature that the content of carbon nanotubes plays a huge role in the flammability properties of the PLA. The PLA/CNT with 1-9 phr content of the CNTs was prepared by the twin-screw extruder. PLA with 3 phr CNT had a rating of V2 from the UL-94, which was due to a low rate of burning in comparison to pristine PLA and PLA/CNT_1phr. However, the higher CNTs contents i.e., 5, 7, and 9 phr contents, were found to have passed the UL-94 with a V0 rating. The observations from the UL-94 showed that these compositions showed slow ignition, with not all of the samples burning up to the holding clamp. Furthermore, there was no melt-dripping for all samples, with cotton remaining in contact with the flame process. Layered molybdenum disulfide (MoS_2) functionalized with the metal oxide (M_xO_y) was utilized as a catalyst for the growth of carbon nanotubes by a chemical vapour-deposition (CVD) method [24]. The produced MoS_2/M_xO_y/CNT was used as a flame-retardant filler to enhance the flame retardancy of the PLA polymer matrix. The composites were fabricated by a twin-screw extruder with 0.5 to 2wt%. The fire retardancy of the composites was analysed by the cone calorimeter. According to Figure 5.12, MoS_2/Fe_2O_3/CNT flame-retardant filler showed the highest decrease in CO emission when compared with neat PLA and other flame-retardant filler(s) (i.e., MoS_2/Co_2O_3/CNT, and MoS_2/Ni_2O_3/CNT)-reinforced PLA composites. This was ascribed to the formation of a thermally stable char barrier, which is able to facilitate the protection of the PLA from heat and block the combustible gases from moving out of the system, as gases have a tendency of acting as a catalyst during a process of burning. All the other parameters of the flammability studies such as heat release rate (HRR), total heat release (THR) and HRC were also reported to have decreased with the addition of the flame-retardant fillers. As extensively discussed in this chapter, it is evident that the incorporation of the flame-retardant fillers blocked the oxygen from entering the system and combustible gases from leaving the system, as a result enhancing flame retardancy. In summary, the addition of few layered MoS_2 had the ability to reduce the CO by 70% when compared with PLA, while the incorporation of $MoS_2/Fe_2O_3/CNT$ reduced CO by 90%. The conclusion from the cone calorimeter is such that the MoS_2 and $MoS_2/Fe_2O_3/CNT$ (Table 5.2) were better flame retardants, while the $MoS_2/Co_2O_3/CNT$, and $MoS_2/Ni_2O_3/CNT$ displayed instability in terms of the barrier at the composition 0.5wt% and 1wt%.

Table 5.2. Summarized results from the MCC of all investigated samples [24]

	FR load [wt%]	HRC [Jg^{-1}K^{-1}]	pHRR [Wg^{-1}]	THR [kJg^{-1}]
PLA	—	703.6	558.3	22.1
PLA MoS_2	0.5	555.2	395.9	16.6
	1	618.9	476.3	18.8
	2	602.6	468.6	17.9
PLA $MoS_2/Co_2O_3/CNT$	0.5	560.2	337.1	21.6
	1	571.3	332.1	20.4
	2	565.4	318.7	19.1
PLA $MoS_2/Fe_2O_3/CNT$	0.5	621.6	483.7	19.6
	1	607.2	471.6	18.7
	2	595.9	449.5	17.9
PLA $MoS_2/Ni_2O_3/CNT$	0.5	550.2	375.2	21.1
	1	548.2	363.3	20.2
	2	543.3	330.8	19.8

Phosphate(s) have been used in the past as flame-retardant materials for polymer matrices. Flame-retardant composite-based on PLA fabricated by melt-blending in combination with carbon nanotubes (MWCNT) and Tri(1-hydroxyethyl-3-methyimidazolium chloride) phosphate (IP) modified MWCNT (MIP) was reported by Hu et al [27]. The composites with 5wt% of the MWCNT, IP, and MIP-reinforced PLA was prepared by direct melt-compounding 170°C at 60 rpm for a duration of 8 minutes [27]. The flammability properties were analysed by cone calorimeter, which is a bench-scale fire-testing technique. The neat PLA was found to have an HRR peak around 324 kW/m^2, which was higher than PLA/5MWCNT, PLA/5IP and PLA/5 MIP composites. The addition of the MWCNT into the PLA reduced the HRR peak from 324 kW/m^2 to 176 kW/m^2, while the incorporation of 5 IP and 5 MIP reduced the HRR by 155 and 240 kW/m^2, respectively. Amongst the three flame-retardant fillers, MIP showed a weak neatwork structure, which resulted in poorer flame retardancy. Similarly, with the other parameters such as total heat release (THR), the THR value of PLA was found to be 49 MJ/m^2, while the THR of PLA/5 MWCNT and PLA/5 MIP were found to be 47 MJ/m^2 and 34 MJ/m^2, respectively. Better flame resistance in the presence of MWCNT and IP was ascribed to an effective char layer that is able to protect the matrix from heat and hence prevents further burning of the matrix.

5.4. Future Recommendations

One of the major disadvantages associated with PLA is its ease of flammability. Ease of ignitibility is associated with high flaming drips, which may result in a quick spread of flame. Carbon-based materials have the ability to improve the flame resistance of PLA as discussed in this chapter, and their synergy with other fillers was proven to have enhanced the flame resistance more than single fillers. There is a need to employ a synergy of the carbon-based filler with the nitrogen-containing flame-retardant filler-reinforced PLA composites. The reason why this kind of flame-retardant fillers are suggested is due to their low toxicity, less smoke, effective flame resistance, and little or low corrosiveness [28]. Most importantly, nitrogen flame retardants have the ability to produce non-flammable gases in the combustion reaction(s) and as a result improve the flame retardancy.

5.5. Conclusion

Poly(lactic acid) has various advantages such as environmental friendliness, recyclability and biocompatibility. Besides the advantages, there are various drawbacks that are associated with PLA, such as poor flame resistance. Based on the above statement, PLA has to be reinforced with flame-retardant fillers in order for it to replace conventional polymers. Carbon-based fillers and their derivatives appear as favourable candidates to enhance its flammability resistance. The most-studied flame-retardant fillers which belong to the family of carbon-based fillers include graphite, together with its derivatives and carbon nanotubes. Various factors were found to play a key role in the flammability of the PLA/carbon-based fillers, factors such as the content of the flame-retardant filler(s), synergy with other flame-retardant fillers and the modification of the flame retardant. Generally, the incorporation of the carbon-based fillers enhanced the flammability resistance of the PLA. This is due to their ability to form a protective char. A synergy of the carbon-based fillers with other flame-retardant fillers was observed to have enhanced the flame resistance more than the carbon-based fillers alone. A combination of fillers has a tendency to form a thick, compact, and continuous char layer, which is more effective in protecting the matrix against heat and also blocking the flammable volatile materials from leaving the system.

References

[1] Sbardella F, Martinelli A, Di Lisio V, Bavasso I, Russo P, Tirillò J, Sarasini F. Surface modification of basalt fibres with ZnO nanorods and its effect on thermal and mechanical properties of PLA-based composites. *Biomolecules* 2021; 11:200 (1-19).

[2] Chatiras N, Georgiopoulos P, Christopoulos A, Kontou E. 2019. Thermomechanical characterization of basalt fiber reinforced biodegradable polymers. *Polymer Composites* 2019; 40:4340–4350.

[3] Eselini N, Tirkes S, Akar AO, Tayfun U. Production and characterization of poly (lactic acid)-based biocomposites filled with basalt fiber and flax fiber hybrid. *Journal of Elastomers and Plastics* 2020;52, 701–716.

[4] Wang X, Wang S, Wang W, Li H, Liu X, Gu X, Bourbigot S, Wang Z, Sun J, Zhang S. The flammability and mechanical properties of poly(lactic acid) composites containing Ni-MOF nanosheets with polyhydroxy groups. *Composites Part B: Engineering* 2020; 183:107568

[5] Barczewski M, Mysiukiewicz O, Matykiewicz D, Kloziński A, Andrzejewski J, Piasecki A. Synergistic effect of different basalt fillers and annealing on the structure and properties of polylactide composites. *Polymer Testing* 2020; 89:106628.

[6] Luo X, He M, Guo JB, Wu B. Flame retardancy and mechanical properties of brominated flame retardant for long glass fiber reinforced polypropylene composites. *Advanced Materials Research* 2023;750-752:85-89.

[7] Koch C, Nachev M, Klein J, Köster D, Schmitz OJ, Schmidt TC, Sures B. Degradation of the polymeric brominated flame-retardant polymeric FR by heat and UV exposure. *Environmental Science & Technology* 2019; 53(3):1453-1462.

[8] Elwafa Abdallah M.A. Brominated flame retardants: current status and future perspectives. *International Scholarly Research Notices* 2014; Volume 2014: Article ID 651834 (1-21).

[9] Shen J, Lian J, Lin X, Lin H, Yu J, Wang S. The flame-retardant mechanisms and preparation of polymer composites and their potential application in construction engineering. *Polymers* 2020; 14(1):82 (1-23).

[10] Tomiak F, Rathberger K, Schöffel A, Drummer D. Expandable graphite for flame retardant PA6 applications. *Polymers* 2021; 13(16):2733 (1-16).

[11] Murariu M, Dechief AL, Bonnaud L, Paint Y, Gallos A, Fontaine G, Bourbigot S, Dubois P. The production and properties of polylactide composites filled with expanded graphite. *Polymer Degradation and Stability* 2010; 95:889-900.

[12] Murariu M, Bonnaud L, Yoann P, Fontaine G, Bourbigot S, Dubois P. New trends in polylactic acid (PLA)-based materials: "Green" PLA -calcium sulfate (nano) composites tailored with flame retardant properties. *Polymer Degradation and Stability* 2010; 95(3):374-381.

[13] Castrovinci A. and Camino, G. Fire-retardant mechanisms in polymer nano composite materials. In: Duquesne S, Magniez C, Camino G. editors. *Multifunctional barriers for flexible structure-textile, series: leather and paper,* vol 97. Berlin, Heidelberg: Springer;2007. p87-108.

[14] Wang D. Wilkie C.A. Fire properties of polymer nanocomposites. In: Mouritz AP, Gibson AG, editors. *Fire properties of polymer composite materials series: solid mechanics and its applications.* Netherlands: Springer; 2006. 143: p.287-312.

[15] Gilman JW. Flammability and thermal stability studies of polymer layered-silicate (clay) nanocomposites. *Applied Clay Science* 1999; 15:31-49.

[16] Babrauskas V. The early history of the cone calorimeter. *Fire Science and Technology* 2022;41(1):21-31.

[17] Melillo JMA, Pereira LM, Mottin AC, Da Silva Araujo FG. Modification of poly(lactic acid) filament with expandable graphite for additive manufacturing using fused filament fabrication (FFF): effect on thermal and mechanical properties. *Polimeros* 2021; 31(2): e2021024 (1-9).

[18] Wu H, Ortiz R, Correa RDA, Krifa M, Koo J.H. Self-extinguishing and non-drip flame retardant polyamide 6 nanocomposite: mechanical, thermal, and combustion behavior. *Flame Retardancy and Thermal Stability of Materials* 2018; 1:1-13.

[19] Fukushima K, Marariu M, Camino G, Dubios P. Effect of the expanded graphite/layered-silicate clay on thermal, mechanical and fire retardant properties of poly(lactic acid). *Polymer Degradation and Stability* 2010; 95:1063-1076.

[20] Zhu H, Zhu Q, Li J, Tao K, Xue L, Yan Q. Synergistic effect between expandable graphite and ammonium polyphosphate on flame retarded polylactide. *Polymer Degradation and Stability* 2011; 96(2):183-189.

[21] Tang G, Zhang R, Wang X, Wang B, Song L, Hu Y, Gong X. Enhancement of flame-retardant performance of bio-based polylactic acid composites with the incorporation of aluminium hypophosphite and expanded graphite. *Journal of Macromolecular Science Part A: Pure and Applied Chemistry* 2013; 50(2):255-269.

[22] Yang P, Wu H, Yang F, Yang J, Wang R, Zhu Z. A novel self-assembled graphene-based flame retardant: synthesis and flame retardant performance in PLA. *Polymers* 2021; 13:4216.

[23] Jing J, Zhang Y, Tang X, Li X, Peng M, Fang, Z. Combination of a bio-based polyphosphonate and modified graphene oxide toward superior flame retardant polylactic acid. *RSC advances* 2018; 8:4304-4313.

[24] Homa P, Wenelska K, Mijowska E. Enhanced thermal properties of poly(lactic acid)/MoS_2/carbon nanotubes composites. *Scientific Reports* 2020; 10:740(1-11).

[25] Guo C, Xin F, Zhai C, Chen Y. Flammability and thermal properties of modified carbon nanotubes in poly(lactic acid). *Journal of Thermoplastic Composite Materials* 2019; 32(8):1107-1122.

[26] Desa MSZM, Hassan A, Arsad A. The influence of carbon nanotubes contents on electrical and flammability properties of poly(lactic acid)/multiwalled carbon nanotubes nanocomposites. *Solid State Phenomena* 2017; 268:365-369.

[27] Hu Y, Xu P, Gui H, Wang X, Ding Y. Effect of imidazolium phosphate and multiwalled carbon nanotubes on thermal stability and flame retardancy of polylactide. *Composites Part A: Applied Science and Manufacturing* 2015; 77:147-153.

[28] Shen J, Liang J, Lin X, Lin H, Yu J, Wang S. The flame-retardant mechanisms and preparation of polymer composites and their potential application in construction engineering. *Polymers* 2022;14(1):82.

Chapter 6

Mechanical Properties of the PLA/ Carbon Based Fillers Composites

Abstract

Carbon materials-filled polylactic acid composites attract significant attention due to their superior distinctive properties. These properties are essentially dependent on several factors such as filler-type, filler-content, and filler-modification, which affect their physical and chemical interaction with PLA. The inclusion of a secondary filler to overcome the issues associated with single carbon-filled PLA also affects the resultant properties. In this chapter, the mechanical properties of the PLA/carbon-based fillers are elucidated. The effect of filler modification, type of carbon-based filler, and the inclusion of the secondary filler on the mechanical properties of PLA are also discussed.

Keywords: mechanical properties, filler-modification, 3D printing, tensile strength, impact strength

6.1. Introduction

Over the past decades, the interest in the use of biopolymers as replacement for conventional petroleum-based polymers has continued to increase [1-3]. This is exemplified by the volume of research outputs and patents produced annually. The main reason for such great interest is their unique properties, such as biodegradability, renewability, and abundance [1-5]. Other distinctive properties include biocompatibility and ease of modification, making these polymers useful for a vast number of applications, such as packaging, tissue engineering, drug delivery, medical implants, and wound dressing [6, 7].

Biopolymers are polymers synthesized chemically from natural sources [1, 4]. Furthermore, live organisms have the capability to biosynthesize these polymers directly under appropriate conditions [1]. However, biopolymers have certain drawbacks that limit their usage in advanced applications, including poor processability and poor mechanical performance. These issues can be resolved by reinforcing these polymers with various (nano)fillers. The resulting 'green' (nano)composite materials can degrade when exposed to several environmental factors [8]. In this case, the biopolymer regulates the structure, environmental durability and tolerance, whereas the (nano)filler governs mechanical performance. Some promising biopolymers for green composites are polylactic acid (PLA), poly-ε-caprolactone (PCL), polybutylene succinate (PBS), polyhydroxyalkanoates (PHA), natural rubber, and chitosan [1, 9-11]. Amongst these biopolymers, a great deal of attention has been given to PLA due to its properties similar to conventional petroleum-based polymers. It is biocompatible, biodegradable and a renewable polymer. PLA is widely used in various sectors, such packaging, agriculture, personal care, and health. The inclusion of different fillers, such as clay, plant fibres, and carbon (nano)fillers were found to improve the overall properties of PLA [10-13]. The major issue associated with PLA is its poor toughness and conductivity, which limit its broad application. The incorporation of carbon-based fillers has been reported to overcome the issues related to the conductivity of PLA. It is recognized that the inclusion of these fillers further enhances the brittleness on the resultant (nano)composites. Copolymerization, co-blending and plasticization of PLA have been reported to resolve the brittleness of the resultant composites [14]. The aim of this overview is to discuss the influence of carbon fillers on the mechanical properties of PLA. In addition, the effect of inclusion of the carbon fillers into PLA-based polymeric materials (e.g., plasticized, blended and copolymerized PLA) is also covered. The preparation processes for manufacturing PLA-based materials affect the mechanical behaviour of the resulting PLA/CBF (nano)composites [15, 16]. For instance, Rane et al. [16] studied the effect of preparation method on the mechanical properties of carbon black (CB)-filled PLA. The authors prepared the composites using two methods, i.e., solution-casting and melt-mixing, followed by melt-compounding. The composite prepared using melt-mixing followed by melt-pressing had higher tensile strength than those prepared using solution casting. The impact strength was also higher for samples prepared through melt-mixing followed by melt-pressing. The superior mechanical properties for both processing methods were obtained at 2.5%CB loading due to the formation of a CB-CB network within PLA and strong interfacial adhesion, which contribute to fracture resistance. This demonstrates that the processing method and the filler content have to be considered carefully to achieve the anticipated mechanical properties. Moreover, carbon (nano)fillers' mechanical properties are different, which lead to different mechanical performances [15].

Similar to traditional composite materials, the interaction between PLA and carbon (nano)fillers and their dispersion play a crucial role in the resultant mechanical properties. In most cases, strong interfacial adhesion and homogeneous dispersion of the fillers are preferable to improve the overall mechanical performance of PLA. Such morphologies are realized by their good stress transfer from PLA to the filler and thus improve mechanical and thermomechanical properties. In summary, the mechanical properties of PLA/carbon (nano)fillers are affected by processing parameters, filler-type, polymer-filler interaction, and post-modification of polymer composites. Therefore, all these factors are important to obtain the anticipated mechanical properties. This chapter discusses all the factors that influence the mechanical properties of PLA/CBF composites. It also sheds light on the practices that are frequently used to enhance the mechanical performance of PLA/CBF composites.

6.2. Factors Affecting Mechanical Properties of PLA

Similar to any thermoplastic polymer, the mechanical properties of PLA are strongly affected by molecular weight and the degree of crystallinity [13, 17-19]. In addition, the crystal form and crystalline morphology of PLA influence its mechanical properties [13, 17]. PLA crystallize into three different forms, i.e., α, β and γ [13, 18, 20, 21]. The most common and stable crystallization form is the α form. High temperatures and stretching of the α form lead to a formation of the β form; meanwhile γ can be formed by epitaxial crystallization [18, 20, 22]. The crystallized PLA at a low temperature results in the α' form [17, 21]. Since the α' is less packed and highly disordered, the PLA containing higher α' turned out to improve the overall toughness of PLA. However, a higher tensile modulus and lower elongation-at-break are obtained for PLA with a higher content of α form than α' form. The α form is highly ordered with a densely packed crystal structure. Therefore, the crystal composition and geometry are essential to be tailored in order to achieve the anticipated tensile properties.

PLA is obtained from its building block known as lactide or lactic acid. Lactic acid can be produced chemically or biologically into isomeric forms, i.e., D-lactic acid and L-lactic acid or race-mixture of D- and L-lactic acid.

The ratio of L- and D-lactic isomers are also crucial for the mechanical performance of PLA, because they control the crystallinity of the polymer. Using L-lactic acid leads to the formation of highly crystalline poly(L-lactic acid) (PLLA) when the L-isomer is above 90%. On the other hand, the D-lactic acid leads to the formation of poly(D-lactic acid) (PDLA). A formation of a stereo-complex between these homopolymers significantly improves the overall tensile properties and elevates thermal stability. The molecular weight of PLA often depends on the polymerization route. Direct polycondensation is usually employed when a low molecular weight PLA is required. The direct polymerization method (i.e., step growth polymerization) involves the removal of water as byproduct. This process makes it difficult to achieve high molar mass PLA and thus low molecular weight can be obtained (>50, 000 g mol^{-1}). Therefore, PLA obtained using this method possesses poor mechanical properties. On the other hand, azeotropic dehydration condensation and lactide ring-opening polymerization lead to high molar-mass PLA. Nonetheless, PLA has a fairly high mechanical strength, an acceptable flexural modulus up to 140 MPa, and a tensile modulus ranging between 5-10 GPa (Table 6.1). The modification of PLA has been the core focal point of the academic communities and industries in order to develop materials that can compete with conventional polymers in some industries, such as packaging and healthcare. Some of the fillers that are often used as reinforcement of PLA are carbon-based (nano)fillers. In this regard, the interfacial adhesion between the composite component, filler content, and filler dispersion are reported to be important factors that control the resulting mechanical performance of the PLA/CBF composites [23-25]. These factors often depend on the preparation method, type of filler, surface modification of either the filler and/or the host polymeric material [23, 24]. Table 6.1 shows an overview of the mechanical properties of PLA/CBFs.

6.3. Mechanical Properties of PLA-CBF Composites

The mechanical properties of reinforced polymer composites are directly dependent on factors such as filler-type, loading, size, shape, and orientation of the filler (Table 6.1) [26]. These factors are strongly affected by the strength of the interaction between the phases, extent of dispersion, and the amount of particle agglomeration. Preparation methods essentially play a critical role in the extent of filler dispersion and interfacial adhesion between the (nano)filler and host polymeric materials. Solution mixing is one of the methods that afford fairly dispersed filler morphology [23, 25, 27, 28]. It has been the most-used method to develop CBF/PLA composites. The modification of the CBFs has been the topical subject in order to obtain a highly dispersed CBFs phase within PLA. In addition, these modifications improve the interfacial adhesion between CBFs and PLA, and thus better stress transfer from PLA to CBF. The latter is justified by improved tensile strength and elongation-at-break when compared to both neat PLA and unmodified CBFs-based composites. Generally, tensile modulus was found to increase with an increase in CBF content due to the reinforcing effect of CBFs. The modification of the CBFs further improves the tensile modulus of the composites.

6.3.1. Solution Mixing

A wide variety of carbon-based (nano)fillers, such as carbon black (CB) [16, 26, 29], carbon nanotubes (CNTs) [23], graphene-based fillers, and carbon fibres, were incorporated into PLA using a solution-casting method. These fillers were found to improve the resulting mechanical properties depending on a number of factors (Table 6.1). Several researchers reported on the incorporation of CNTs into PLA using a solution mixing technique [23]. Chloroform, THF and DMF are commonly used solvents for such purposes [23, 25]. In most cases, CNTs tend to agglomerate due to strong Van der Waal forces that lead to poor stress transfer between the polymer and the CNTs. In this case, fillers act as stress centres under tensile stress. This allows for the propagation of the crazing; hence resulting in poor tensile strength and drawability. The modification of CNTs before incorporation into PLA enhances interfacial adhesion between CNTs and PLA which, in turn, improves stress transfer between the composite component [23]. In addition, these kinds of modification often improve the dispersion of CNTs within PLA, hence improvements on the overall properties of the resulting composites. The purification of CNTs using a mixture of sulphuric acid and nitric acid is often employed to introduce carboxyl groups onto CNTs [25, 30]. This treatment also reduces the length of the CNTs [25]. The presence of COOH-CNTs was reported to enhance the mechanical properties of PLA because of the strong interaction resulting from covalent/hydrogen bonding between –COOH and –OH groups [25].

Table 6.1. Mechanical properties of PLA/CBF composites prepared using solution casting

Formulation	Filler Content (%)	Chemical modification	Post-process	Tensile modulus (GPa)	Tensile strength (MPa)	Elongation-at-break (%)	Impact strength (MJ m^{-3})	Comments	Refs.
PLA	-	-	-	0.5-4.5	50-59	7-10	2.7-15	Besides the tensile testing parameters, the properties of PLA control the resulting mechanical performance	[19, 25, 27, 31, 32]
PLA/CNTs	2.5	-	Solution casting	0.7	57	10	-	Only tensile modulus increased while other remained similar to neat PLA	[25]
	0.5-2.5	Oxidation using acid mixture (CNTs-COOH)		0.6-1	58-74	19-22	-	Tensile strength, tensile modulus and elongation-at-break increased with CNTs content because of strong interfacial adhesion of the composite components due to CNTs modification	
PLA	-	-	Compression moulding	2.1 0.1	55.2 0.6	7.3 0.9	-	Grafting PLA onto CNTs resulted in better dispersion and interaction between CNTs and PLA, which afforded better mechanical performance	[27]
PLA/CNTs	2	-		3.5 0.2	59.8 1.1	6.8 0.6	-		
PLA/PLA-g-CNTs	2	Oxidation using acid mixture (CNTs-COOH)-Acyl chloride (CNTs-OCl)-glycol (CNTs-OH)-functionalization-PLLA grafting		4.3 0.1	64.7 0.9	12.4 1.2	-		
PLA-g-	0.3-5	PLLA-grafting	-	-	60	-	-	3% loading of PLA-g-GO were	[31]

Formulation	Filler Content (%)	Chemical modification	Post-process	Tensile modulus (GPa)	Tensile strength (MPa)	Elongation-at-break (%)	Impact strength (MJ m^{-3})	Comments	Refs.
GO/PLA									
PLA	-	-	Oven dried	4.5 ± 0.034	-	-	-	found to be optimal to obtain maximum tensile strength	[32]
PLA/GO	0.1-1a	-		4.7 ± 0.045	-	-	-	Surface modification and hybridization of the fillers improve mechanical properties of PLA when compared to unmodified-based fillers	
PLA/GO/GO-g-POSS		Polyhedral oligomeric silsesquioxanes		5.8 ±0.053	-	-	-		
PLA/GO-g-POSS				6.0 ± 0.028	-	-	-		
PLA/GO-octadecylamine (OD)	0.1-1.0	Alkylation using octadecylamine	Vacuum drying	2.2-3.0	50-61	6.5-15	142-687	Maximum tensile strength and modulus of 60.7 and 3 GPa were achieved with the incorporation of 0.2% of modified GNPs and elongation-at-break, and impact strength reached 15% and 687MJm^{-3} at 0.6 GNPs loading	[28]
PLA	-	-	-	2.83	36.43	8.5	-	Elongation-at-break was significantly improved at the expense of tensile strength and modulus	[26]
PLA/thymol 95/5	-	-		2.02	24.12	72	-		
PLA/CB/thymol	0.5-10	-		1.24-0.76	25.5-14	90-320	-		

a=gram

Tensile strength and tensile modulus were reported to increase with COOH-CNTs content [25]. In addition, elongation-at-break was also found to increase from 10% to ~20%, regardless of the content [25]. The improved mechanical performance was ascribed to the good reinforcing effect of CNTs as well as good adhesion between CNTs and PLA. In their study, Yoon et al. [23] functionalized COOH-CNTs using direct melt-condensation of L-lactide monomer. The authors purified CNTs using a mixture of H_2SO_4 (1mol) and HNO_3 (3 mol) through sonication for an hour, followed by reflux at 120°C for 2 hours to introduce –COOH groups onto CNTs (COOH-CNTs) for grafting of PLA onto CNTs via a melt-polycondensation process. PLA-grafted-CNTs (PLA-g-CNTs) were mixed with virgin PLA using chloroform as a solvent, followed by melt-pressing to afford films with ~0.2 mm thickness for mechanical characterization. In comparison with virgin PLA, the CNT- and COOH-CNT-based PLA composites exhibited a slight increase in tensile strength and tensile modulus. A significant increase in tensile strength and modulus with only 1% of CNT-g-PLA into PLA was noted due to strong interfacial adhesion and good dispersion of CNTs. The increments of 32% and 47% for tensile modulus and strength, respectively, were obtained with the inclusion of 1% PLA-g-CNTs, when compared to the virgin PLA. These factors improve stress transfer between composite components. Since the content of the filler plays an essential role in the mechanical performance of the composites, the authors report that tensile strength and modulus increased with an increase in filler content from 0.1 to 1.0%; after that a drastic decrease was observed. Even when using a comparison between experimental and calculated modulus it was noticed that experimental moduli were higher than the calculated moduli at fairly low filler content and the opposite prevails at higher filler content. This was attributed to the better dispersion at low content; meanwhile aggregates are more dominant at higher filler content. The chain length of PLA has a significant influence on the dispersion state of the CNTs which, in turn, affect the mechanical performance of the resulting composite materials [24]. Elsewhere, the effect of chain length of PLA on the dispersion of CNTs was studied using a solution mixing technique [24]. It was reported that purified CNTs using acid treatment were used to graft PLA with different chain length by in-situ ring-opening polymerization (ROP). It was noticed that the tensile strength and tensile modulus increased in the presence of purified CNTs to a certain extent, i.e., from ~1928 to 2009 MPa and ~49.3 to 53.6 MPa. This is ascribed to the limited interaction between PLA and CNTs as well as the rigidity of the filler. In the case of PLA-g-CNTs it was reported that both the tensile strength and tensile modulus further increased with the chain length of PLA grafted onto CNTs. The tensile modulus was found to increase linearly with the chain length (molecular weight) grafted onto CNTs. The tensile strength, however, increased until the molecular weight of ~200 g mol^{-1} of grafted PLA, whereafter it reached a plateau. This was attributed to strong interfacial adhesion between CNTs and PLA and better dispersion due to PLA being grafted onto CNTs. Using a model assuming CNTs were randomly oriented with the matrix, it was found that the calculated moduli underestimated the moduli for CNTs grafted with a molecular weight between 145 and 287 g mol^{-1} due to better interfacial adhesion-promoting stress transfer from PLA to incorporated PLA.

Dynamic mechanical properties of PLA/COOH-CNTs were reported by Chrissafis, et al. [25]. PLA storage modulus was found to be constant up to 45°C, i.e., close to the glass transition temperature, and then decreased, followed by a slight increment due to cold-crystallization. Similar observations were recorded for the PLA/COOH-CNTs composites. It was noticed that the increase in COOH-CNTs resulted in an increase in storage modulus because of the reinforcing effect of CNTs. From tan δ it was found that the T_g of PLA shifts to higher values with an increase in CNTs content. In addition, the peak area decreased with an increase in CNTs content. These are ascribed to CNTs restricting molecular chain mobility of PLA resulting from good dispersion and strong interaction. The formation of hydrogen bonds between CNTs and PLA further contributed to the molecular chain mobility. Chiu et al. [30] further modified acid-purified CNTs with 3-isocyanatopropyl triethoxysilane (A-CNTs) to improve the overall dispersion and interfacial adhesion between the filler and PLA. It was observed that the presence of CNTs enhanced the storage modulus of PLA throughout the investigated temperatures, regardless of modification. The storage modulus increased with an increase in CNTs content. This was ascribed to the presence of stiff materials that contribute to the reinforcing effect within PLA. These results were justified by an increase in Tg when compared to neat PLA. The presence of A-CNTs improved the storage modulus of PLA more than unmodified and acid-treated CNTs. The Tg increases from ~67°C to ~72°C, and the storage modulus from ~2.225 to ~3720 MPa when compared to acid-treated CNTs-based composites. This is the result of strong interaction between PLA and A-CNTs as well as a better dispersion of CNTs due to surface modification applied.

Similar to CNT-based composites, the mechanical properties of graphene-based composites were found to be dependent on the modification of the CNTs and their concentration [31]. Wang et al. [31] report on the effect of graphene oxide (GO)-grafted-(L-lactide) (GO-g-PLA) on the mechanical properties of the resulting composites. The tensile strength of the GO-g-PLA-based composites was higher than those of neat PLA and the unmodified GO-based composites (Figure 6.1). The unmodified GO-based composites exhibited a tensile strength even lower than neat PLA. This observations were attributed to better interaction between PLA-g-CNTs and PLA via covalent bonds that allowed better interior stress transfer from PLA chains to graphene nanosheets. The optimal content of grafted GO was found to be 3%, with an increment of ~39% when compared to the neat PLA. Nonetheless, the increment in the tensile strength was attributed to good stress transfer from PLA to GO as justified by coarse tensile surface fracture with some particles on the surface of polymer reflecting good interfacial adhesion (Figure 6.1c). In the case of PLA a smooth surface was observed (Figure 6.1c), indicating a brittle fracture of PLA. Meanwhile a PLA/5%GO composite displayed a coarser fracture with some visible aggregates with no individual filler present on the fractured surface (Figure 6.1e). This explains the reduction of tensile strength at a high PLA-g-GO greater than 3% GO loadings. It is worth mentioning that the crystallinity of PLA increased up to 3% loading of PLA-g-GO, which also contributed to an increment in tensile strength. The decrease in crystallinity beyond 3% loading led to agglomerates that hindered the alignment of PLA chains and thus affect the crystallization process negatively. In general, the presence of carbon-based (nano)fillers (CBFs) improves the mechanical properties of PLA to a certain extent using a solution-casting method. The reinforcing effect of CBFs is justified by an increase in tensile modulus and storage modulus with the filler content. The surface modification of the CBFs further improves the mechanical performance of the resulting composite materials due to the better dispersion of the filler and their enhanced compatibility with the host matrix. This is as a result of better stress transfer from PLA to the filler. In addition, the commonly used CBFs modification involves their oxidation to introduce oxygen functionalities that are essential to form chemical interactions with PLA to a certain extent. However, further modification of CBFs after the oxidation process often introduced other functionalities which contribute to homogeneous filler dispersion as well as strong interfacial adhesion between composites' components. It can be argued that these modifications are often expensive and complex, which makes them unfavourable for industrial scale-up. The use of carbon fillers bearing functionalities (e.g., graphene oxide (GO)) that can improve interfacial adhesion and dispersion of the filler is of essence to overcome these issues. In addition, the decoration of the CBFs with other fillers that can impart the interfacial adhesion and dispersion of the fillers could be an alternative solution to resolve some of the issues that involve the dispersion of CBFs within PLA and the anticipated interaction. Since PLA-grafting onto CBFs is the commonly employed method to enhance the dispersion and interfacial adhesion, in situ polymerization of PLA is of interest when it comes to the preparation of composites with high mechanical properties.

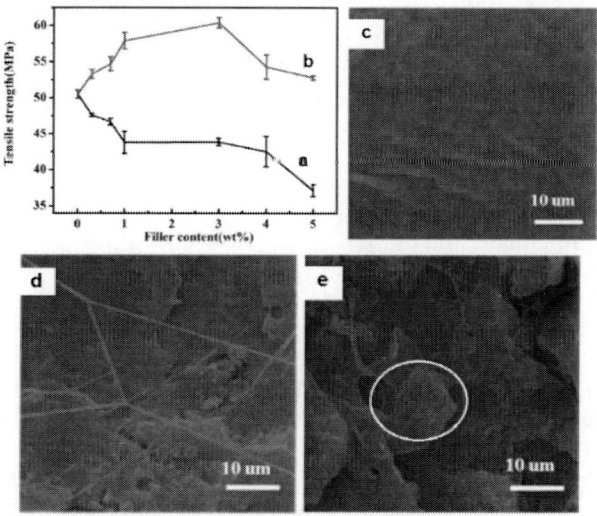

Figure 6.1. Tensile strength of GO-based composites: (a) GO/PLA and (b) GO-g-PLA/PLA. SEM images of (c) neat PLA, (d) 3%GO-g-PLA/PLA and (e) 5%GO-g-PLA/PLA. Reproduced from Wang et al. [31]. Open Access.

In spite the fact that CBFs are often introduced into plasticized PLA to compensate for reduction in tensile strength and modulus, the introduction of carbon black (CB) into PLA plasticized with 5% thymol showed opposite behaviour [26]. In this respect, the content of CB particles was varied between 0.5-10% to study their effect on the mechanical properties of a PLA/thymol blend. It was reported that the tensile strength and Young's modulus were reduced by the presence of thymol as plasticizer, and the inclusion of CB further decreased these parameters. The elongation-at-break was found to increase with an increase in CB content at the expense of tensile modulus and tensile strength. The authors did not provide reasons why is this the case, because one could expect the presence of CB to improve both tensile strength and Young's modulus of PLA. A comparison between solution-casting and melt-mixing using CB as filler for PLA was reported by Rane et al. [16]. Melt-mixing, followed by melt-pressing had better tensile strength and impact strength than solution-casting-prepared composites. The optimal concentration of 2.5% was reported to give the highest mechanical properties with an increase in content to 5%, leading to a reduction in mechanical properties.

6.3.2. Electrospinning

Electrospinning technique afford the fabrication of nanofibers with controllable shapes, sizes, surface-to-volume ratio and porosity [33, 34]. It is one of the simplest up-scalable methods to produce nanofibers with diameters ranging between nano- and micro-meters with a promising application in different fields, such as filtration, textiles, electronics, biosensors, biomedical, and agriculture. The resulting fibres are often fragile, which limits their application in high-end application. Subsequently, various fillers have been introduced into the system to achieve the desired mechanical properties in order to qualify in the anticipated application. The influence of GNPs on the mechanical properties of PLA/poly-ε-caprolactone (PCL) blends was investigated by Chiesa et al. [33]. It was observed that low loadings of GNPs enhanced tensile strength and modulus (Figure 6.2). Maximum tensile strength of ~6.6 MPa was attained with 0.5%GNPs loading. At 2% and 4%GNPs loading the lowest elastic modulus was achieved. Similarly, tensile modulus was the highest for 0.5 and 1% loading, reaching a value of 36.3 and 33.6 MPa. Both neat polymer-based membrane and 2% loading had a comparable tensile modulus, i.e., 23.2 and 20.6, while 4% loading reached a value of 16.5 MPa. Elsewhere, it was reported that tensile modulus and strength of PLA was improved with the introduction of nanographene oxide (nGO) [35]. The increase in nGO content further improved both tensile strength and modulus of the resultant electro-spun composite fibres (Figure 6.2b). This was due to strong interaction between nGO and PLA as well as the homogeneous distribution of nGO within the host matrix. It was discovered that PLA-based fibres broke suddenly, while PLA-nGO fibres broke more gradually and with better alignment (Figure 6.2c). This demonstrates that the mechanical performance of a PLA-CBF composite could be tailored towards its anticipated application. The mechanical properties can also be modified by blending PLA with flexible polymers in the presence of CBFs. For instance, the inclusion of MWCNTs into a PLA/PCL blend resulted in mechanical performance between that of individual polymers. It was pointed out that the increase in MWCNTs from 0.25 to 1.0wt% increased both tensile modulus and strength due to a better dispersion of CNTs, and thus allows load transfer from the polymer to the filler. Beyond 1% loading there was a detrimental effect on the mechanical performance because of agglomeration of the filler. 1% loading into PLA resulted in superior tensile strength and modulus when compared to PCL/CNTs-1%, but comparable to PCL/PLA-CNTs1%. However, elongation break was better in PCL/PLA1% (78% vs. 60% for PLA/CNTs1%).

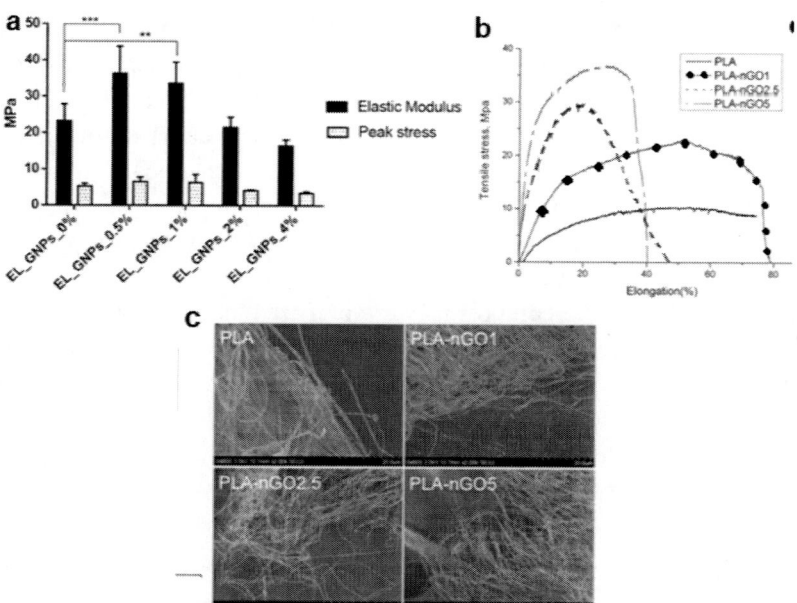

Figure 6.2. (a) Tensile modulus and tensile strength neat polymer-based and composite nanofiber membranes. Reprinted with permission from Chiesa et al. [32]. (b) stress-strain curves and (c) SEM images for electro-spun PLA and PLA-nGO composites. Reproduced with permission from Wu et al. [35].

6.3.3. Melt-Compounding

The mechanical properties of PLA/CBF composites prepared using conventional melt-compounding methods are reviewed in Table 6.2 [36, 37]. The preparation of PLA/CBF using melt-compounding often results in carbon filler agglomeration when introduced into molten PLA matrix. Such agglomerates can affect the mechanical properties of PLA negatively. The surface modification of CBF is the most suitable method to improve dispersion as well as interfacial adhesion. Polymerization of methyl methacrylate (MMA) in the presence of graphene leads to PMMA grafting onto graphene, which affords compatibility between graphene and PLA. PMMA bears ester groups that can easily interact with the PLA ester groups. This process was adopted to avoid complications and complexities involving PLA polymerization as a form of functionalization. The PMMA-functionalized graphene (PFG) was incorporated into PLA using an internal mixer. Tensile strength increased with PFG content from 60 MPa to 65 MPa, whereas elongation-at-breaks decreased from 6% to 4% with the addition of 5% PFG. Tensile modulus was also raised with an increase in PFG loading. This was attributed to the rigidity of PFG. However, the increments in tensile strength were attributed to strength interaction between PLA and PFG as well as uniform dispersion of the filler. The chemical modification, depending on the carbon filler-type, affords a highly dispersed phase and strong interaction, regardless of the preparation method. In their study, Gou et al. [38] report on the effect of unmodified CB content on the mechanical properties of PLA. The composites were prepared using internal melt-mixer followed by compression-moulding. The increase in CB content resulted in an increase in tensile modulus, and tensile strength up to 12% loading. Beyond 12% CB loadings, there was a drastic decrease in both tensile strength and modulus. However, elongation-at-break was found to decrease with an increase in CB loadings. This phenomenon was attributed to better dispersion of CB at loadings below 12%, above which CB agglomerates and acts as stress concentration points, resulting in poor mechanical performance. The decrease in elongation-at-break was ascribed to the reinforcing effect of rigid CB. The cooling of the (nano)composites has a significant impact on the crystallization behaviour of the host polymeric material. As expected, Delgado et al. [39] report that slow cooling rates result in highly crystalline (nano)composites when compared to faster cooling. Tensile modulus and elongation-at-break were not affected by changes in crystallinities through different cooling rates. These properties were not even affected by the content of CB. The (nano)composites exhibited an increase in tensile strength with an increase in CB content for highly crystalline (nano)composites obtained from slow cooling rates. These nanocomposites performed even better than the ones based on faster cooling rates. In most cases, flexible polymers and plasticizers are introduced in PLA-based systems to improve the overall toughness [19, 40-42]. This process often resulted in the deterioration of tensile strength and modulus of PLA [19]. With the addition of carbon-based fillers, mechanical properties of PLA were significantly improved at fairly low concentrations [43]. The influence of plasticizer and CB loadings is reported by Wang et al. [43]. Beside the fact that the presence of plasticizers improved the interaction between CB and PLA, the inclusion of CB into neat PLA resulted in better tensile strength at the expense of elongation-at-break. It was noticed that tensile strength increased with CB loading. The extent of influence on the mechanical properties was directly dependent on the plasticizer, i.e., either acetyl tributyl citrate (ATBC) or poly(1,3-butylene adipate) (PBA). The presence of PBA significantly improved elongation-at-break and tensile strength when compared to ATBC plasticizer at the same CB loading. The reinforcing effect of CB was confirmed by an increase in storage modulus with CB loading, regardless of the plasticizer-type. Compared to CB/PLA composites, the presence of plasticizers decreased storage modulus because the stiffness was reduced by plasticizer. The glass transition temperature (T_g) was drastically reduced in the presence of plasticizers; however, the inclusion of CB increased the intermolecular interaction of PLA and reduced the free volume and elevated T_g of the composites. Similar observations were reported by Yu, et al [44]. The increase in ATBC decreased tensile strength while increasing elongation-at-break. Despite the presence of CB compensating the deteriorated tensile strength, the inclusion of CB in PLA performed better than the composites prepared in the presence of the plasticizer. Nonetheless, tensile strength was found to increase with CB content demonstrating its reinforcing effect. The optimal content of plasticizer was found to be below 30% and above this value multiphase was formed within the system. CB improved storage modulus and T_g of PLA, while the opposite was observed when plasticizer was introduced.

Cheing et al. [40] report on the effect of graphene nanoplatelets (GNPs) into plasticized PLA. In this case, polyethylene glycol was used as plasticizer to improve the toughness of PLA. The elongation-at-break of PLA was significantly improved in the presence of PEG, and further increased with the inclusion of graphene nanoplatelets (GNPs). The improvements in tensile properties were found to be dependent on the amount of filler introduced into the system. The tensile modulus significantly increased from 424 MPa for PLA/PEG to 780 MPa, with only 0.1% loading. It was reported that there was no difference in tensile modulus with an increase in GNPs content. Tensile strength and elongation-at-break increased up to 0.3% GNPs loading and a further increase in GNPs loading showed the opposite. This was attributed to the reinforcing effect of large aspect ratio GNPS filler and a strong interaction between the fillers and PLA in the presence of PEG. The reduction in mechanical performance beyond 0.3% loading was ascribed to irreversible agglomeration resulting from Van der Waals forces between the graphene nanosheets. Similar observations were reported when using EPO as plasticizer of PLA in the presence of GNPS as reinforcing agent [41, 42]. It was found that tensile strength and elongation-at-break increased by ~27% and ~61% with 0.3% GNPs loading. This was ascribed to irreversible agglomeration of the filler at higher loadings, and thus affecting stress transfer from the host polymer to the filler negatively. The modification GNPs through oxidative exfoliation by thermal or oxidation processes and subsequent chemical reduction is often employed to produce reduced GNPs (rGNPs) [45]. These process result in exfoliated individual graphene nanosheets with a higher aspect ratio when compared to untreated graphene particles [45]. Elsewhere, the reinforcing behaviour of GNPs and rGNPs for PLA in the presence of epoxy palm oil (EPO) and PEG as plasticizers was investigated [45]. The presence of rGNPs increased tensile strength by a 3% margin when compared to GNPs-based composites. For PLA/EPO/rGNPs, the increase of 4% was obtained due to better dispersion, and strong interfacial adhesion resulting from a high aspect ratio of rGNPs. In case of tensile modulus, the incorporation of GNPs significantly improved rigidity of the material when compared to rGNPs-based composites. In spite of the reduction in tensile modulus in the presence of plasticizers, the incorporation of GNPs outperformed rGNPs-based composites. Toughness, however, increased by 262%, 54% and 6% for PLA, PLA/EPO and PLA/PEG-based composites when compared to GNPs-filled composites.

The carbon-type filler was reported to play a significant role in the mechanical properties of PLA [46]. In their study, Batakliev et al. [46] report that CNTs-based composites exhibited superior mechanical performance when compared to GNPs-based composites. Tensile strength and elongation-at-break were found to increase from 1.5% up to 6% CNTs loading (Figure 6.3a-b). The better stress transfer due to strong interfacial adhesion resulted from the high-aspect ratio of CNTs when compared to GNPs. Tensile modulus results indicated that CNTs had more reinforcing capabilities when compared to GNPs. It was indicated that both CNTs and GNPs were well-embedded within PLA with a high degree of porosity (Figure 6.3c-d).

Mechanical Properties of the PLA/Carbon Based Fillers Composites 129

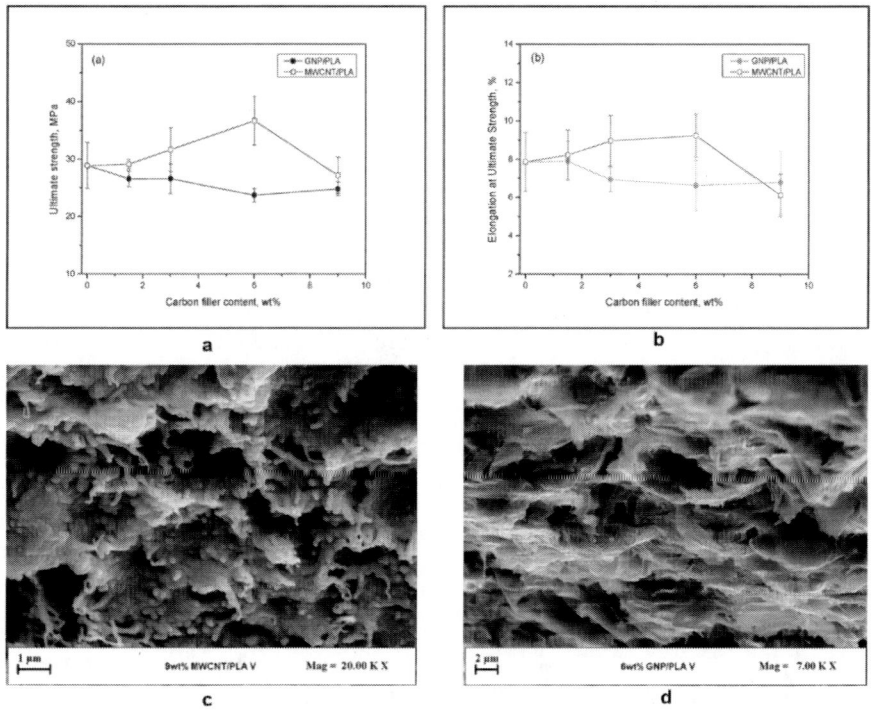

Figure 6.3. (a) Tensile strength, and (b) elongation-at-break for MWCNT/PLA and GNP/PLA (nano)composites. SEM images of (a) PLA/9%CNT, and PLA/6%GNP (nano)composites 6%PLA. Reprinted from Batakliev et al. [46]. Open Access.

6.3.4. Extrusion

A comparison between the size of graphene nanosheets as reinforcement for PLA was carried out by Gao et al. [47]. The authors prepared the composites using GNPs having an aspect ratio of ~500 (GNP-L) and ~2000 (GNP-S) through melt-extrusion, followed by melt-compounding to afford thin films with thickness of 130 µm. The tensile modulus and tensile strength increased with an increase in fillers content; however, the maximum tensile modulus was found to be achieved at different loading, depending on filler size. The optimal content for GNP-S was found to be 7%, while 5% was sufficient for GNS-L. Despite the increase in tensile modulus being associated with the reinforcing effect of the filler, it was deduced that the increase in crystallinity in the presence of GNPs contributed to such increase in Young's modulus. This resulted in larger GNPs particles exhibiting superior tensile modulus when compared to smaller GNPs. This correlated with the increment of crystallinity with GNP-L-based composites increasing by 4-12% and GNP-S by 0.2-12%. The glass-transition temperature, however, was not significantly affected by filler-type and the content, indicating GNPs had no influence on the molecular chain of PLA.

The modification of helical CNTs (HCNTs) with silane afforded their homogeneous dispersion with strong interaction, which are important when it comes to the mechanical properties [48]. The ductility of PLA was improved as justified by an increment of elongation-at-break and impact strength by 205% and 30%, when compared to neat PLA. Encapsulation of CNTs to enhance the toughness of PLA without affecting tensile strength and tensile modulus is reported in the literature [49]. In this case, the silane-treated CNTs were encapsulated by PLA through in situ polymerization. The incorporation of the encapsulated CNTs into PLA using an extrusion method to afford a complete blend containing well-dispersed CNTs. It was reported that 0.5% of encapsulated CNTs were sufficient to improve elongation-at-break, tensile strength and impact strength of PLA significantly. Tensile strength increased by ~40%, elongation-at-break by ~59% and impact strength by ~52%, with the addition of 1% encapsulated CNTs. These results were attributed to chemical bonds between all the layers, which enhanced interfacial adhesion for excellent stress transfer between all composites' components.

6.3.5. Three-Dimensional (3D) Printing

A number of reports focus on impact strengths, tensile and flexural strengths, flexural and tensile moduli of the printed objects when compared to neat PLA. Most of these studies are based on the effect of processing parameters on the mechanical behaviour of the resulting printed components [50, 51].

The printing process of a given design involves different printing parameters that have an influence on the final mechanical properties [51-55]. These include raster orientation, layer thickness, infill level, feed rate, and extrusion temperature [53-57]. It is worth mentioning that the properties of unprinted materials also affect the mechanical properties of the printed products. The incorporation of different fillers and post-treatment of the printed products enhance the overall mechanical performance. In general, the tensile properties are affected when the raster orientation approaches the perpendicular direction of the applied force [53-56]. Cicero et al. [56]

studied the effect of raster orientation s (0/90, 30/-60, and 45/-45) on the tensile and fracture of 3D-printed PLA/graphene (nano)composites. As expected, the presence of graphene increased tensile modulus and Poisson ratio in all orientations, while elongation-at-break decreased. The samples with a raster orientation of 45/-45 exhibited the highest sensitivity to the presence of graphene with modulus increasing with ~44% and elongation-at-break decreased by 42%. The samples produced at 90/0 and 30/-60 tensile modulus increased by 10% and 22%, whereas elongation-at-break decreased by 17% and 16%, respectively. Tensile strength was found to reduce for samples of 90/0; meanwhile there was a 19.2% and 5% increase for 45/-45 and 30/-60 samples, respectively. It was noticed that 30/-60 and 45/-45 exhibited a similar increase in fracture properties. However, 0/90 samples showed an increase in fracture toughness by 13%. Elsewhere, the effect of the printing orientation of PLA/GNPs, i.e., flat, on-edge and up right was investigated [58]. It was reported that flat and on-edge orientations displayed superior tensile and flexural strengths when compared to up-right orientation. The presence of GNPs improved the flexural and tensile strengths for up-right and flat orientations when compared to neat PLA-based materials. These observations were attributed to the enhancement of interlayer adhesion, and the fine-tuning of the 3D-printing can further improve the overall mechanical performance of the printed objects.

The effect of surface modification of CNTs on the mechanical properties of 3D-printed components was studied by Bortoli et al. [59]. CNTs were acid treated using HNO_3 to create defects and introduce surface oxygen functionalities. They found out that functionalization promoted the dispersion of the CNTs within PLA and served as nucleating agents promoting the crystallization of PLA. Tensile strength and tensile modulus increased from ~29.4 MPa to ~41.6 MPa and from 1.18 GPa to 1.48 GPa with the addition of 0.5% of functionalized CNTs, respectively. The incorporation of unmodified CNTs, however, had no influence on the tensile strength (32 ± 5 MPa vs 29.4 ± 0.7 MPa) and tensile modulus (1.18 ± 0.5 GPa vs 1.1 ± 0.1 GPa). Thesis results were explained in terms of better dispersion of functionalized CNTs and their contribution in promoting layer-to-layer adhesion during printing. The 3D printing process was also reported to contribute in improving the tensile strength of the printed components [60]. In this case, it was found that the increase in CNTs loading resulted in an increase in tensile strength. This was attributed to the alignment of the CNTs within PLA during printing.

Table 6.2. Mechanical properties of melt-processed PLA/CBF composites

Formulation	Filler Content (%)	Chemical modification	Post-process	Tensile modulus (GPa)	Tensile strength (MPa)	Elongation-at-break (%)	Impact strength (KJ m^2)	Comments	Refs.
PLA	-	-	Injection moulding	-	65	13	3.4	Surface modification improved the dispersion and interfacial adhesion resulting in a tougher composite material	[48]
PLA/HCNTs	0.2-2.0	CNTs-COOH		-	63-66	21-22	3.8-5.0		
PLA/HCNTs		Silane-treated CNTs		-	68-72	20-40	3.8-5.0		
PLA/CNTs				-	65-67	20-25	3.9-4.7		
PLA	-	-		-	65	62.5	31	Silane treated CNTS and encapsulated silane treated CNTs with PLA resulted in homogeneous distribution with strong interaction that led to improved overall mechanical properties	[49]
PLA/CNTs	1	-		-	70	66	3.5		
PLA/CNTs	1	Silane-treated CNTs		-	74	73	4.2		
PLA/CNTs	0.5-4	Silane –PLA grafted CNTs		-	86-95	80-95	4.2-4.8		
PLA	-	-		-	39.5 ± 0.2	22.5 ± 1.2	15.5 ± 0.2	Better mechanical properties resulted from better dispersion and interfacial adhesion in the presence of compatibilizer	[61]
PLA/PBAT/CNTs	0.1-2	Ethylene-butyl acrylate-glycidyl methacrylate (E-BA-GMA) compatibilizer	Melt-pressed	-	40.5-42.8	20.5-33.6	8.8-27.7		
PLA/PBS/CB	1-2	-	Melt-pressed	0.7-1.0	33-39	-	-	Tensile modulus and tensile strength increased with CB content due to its reinforcing effect	[36]
PLA	-	-	-	5.02 0.53	37.2 0.4	6.0 0.4	-	Epoxy-impregnated CF showed superior mechanical performance due to epoxy improving interaction with the host polymeric material	[37]
PLA/carbon fibres (CF)	10-20	-	-	5.9-6.3	39.9-47.8	3.4-4.2	-		
PLA/ CF		Epoxy-impregnated CF	-	3.8-4.2	43.1-43.6	8.8-9.2	-		

Table 6.3. Mechanical properties of 3D printing materials and printed components

Formulation	Content	Modification	Form	Tensile modulus	Tensile strength	Elongation-at-break (%)	Flexural strength (MPa)	Flexural modulus (MPa)	Impact strength (Jm-1)	Comments	Refs.
PLA	-	-	Filament	0.2-0.4	15.5-72.2	0.5-9.2	52-115	2392-4930	27-192	The properties are directly associated with the grade and processing parameters	[53, 54, 57]
PLA	-	-	Filament		58-72	5-7.4	-	-	1216-1767	The infill pattern influences the mechanical properties because it controls the interaction between the deposited layers	[54]
Smartfil PLA	-	-	Filament	0.3	35.6 ± 3.8	4.2 ± 0.2	85.2± 2.2	2378± 57	29.2 ± 2.3		
Smartfil PLA 3D850	-	-	Filament	0.4	53.4 ±2.1	4.4± 0.3	98.4 ± 3.1	2404± 42	34.6± 3.3		
PLA-graphene	-	-	Filament	0.4	66.8 ± 1.3	2.6± 0.1	98.5 ±2.4	2450± 94	4.4 ± 2.9		
PLA/GNPs	-	-	Printed specimens flat	3.6 ±0.1	63.2 ± 2.8	-	83.4 ±0.5	2142± 10	34.3 ±1.3	On-edge and flat orientations exhibited highest mechanical properties when compared to up-right orientation	[58]
			On-edge	3.6 ±0.9	61.8 ±1.8	-	94.3 ±1.5	2435 ±20	22.5± 1.5		
			Upright	3.6 ±0.1	39.3± 1.1	-	71.1± 8.0	2427± ±40	19.5 ± 1.2		
PLA	-	-	Filament		26	-	-	-	-	Printed components had higher tensile strength that the feeding filaments	[62]
			Printed (45/-45)		31						
PLA/graphene-carbon fibre	1/1		Filament		31						
			Printed (45/-45)		63						

Table 6.3. (Continued).

Formulation	Content	Modification	Form	Tensile modulus	Tensile strength	Elongation-at-break (%)	Flexural strength (MPa)	Flexural modulus (MPa)	Impact strength (Jm-1)	Comments	Refs.
PLA	-	-	90/0	1.18± 0.05	32 ± 5	-	-	-	-	Modification of CNTs improved interaction and dispersion of CNTs	[59]
PLA/CNTs	0.5	-		1.1 ± 0.1	29.4 ± 0.7						
		Oxidation		1.48 ± 0.08	41.6 ± 1.4						

Besides the fact that GNPs are much cheaper than CNTs, CNTs-based 3D-printed materials were also prepared to afford conductive printed components (Table 6.3) [59, 63]. The comparison on the modification of CNTs using HNO_3 on the mechanical properties of the printed components was investigated by Bortoli et al. [59]. The oxidation of CNTs using acid affords the creation of defects and oxygen functional groups that promote interaction between CNTs and PLA, which allows the particles to be well-embedded within PLA. The presence of unmodified CNTs had no significant influence on tensile strength and modulus of PLA. This behaviour was attributed to poor interfacial adhesion between the fillers and PLA, and the poor dispersion of the unmodified CNTs within the matrix. The modified CNTs significantly improved tensile strength and modulus when compared to both PLA and PLA/unmodified CNTs composites. This demonstrates that the modification of CNTs was essential to promote their dispersion within the matrix. DMA results demonstrated that unmodified CNTs improved storage modulus by ~15% at 37°C indicating the reinforcing effect of the CNTs. the peak on the storage modulus curve before Tg was observed due to rigidity corresponding to thermal shrinkage accumulated during processing. The addition of modified CNTs reduced the intensity of this peak and showed a 43% increase in storage modulus when compared to PLA/unmodified composites. This justified the tensile properties of these composites resulting from strong interaction and better dispersion of modified CNTs.

6.3.6. Melt-Spinning

Melt-spinning serves as secondary processing technique that affords the production of composite fibres with diameters as small as a few microns [64, 65]. It is the most economical processing technique when compared to electrospinning, because there are no solvents involved [64, 65]. As explained earlier in Chapter 3, this technique involves the extrusion of molten polymer or composite through spinneret under certain pressure [64]. The extruded material is then quenched in cold air to produce filaments. The resulting filaments often exhibit poor mechanical performance and thus the modification of the fibres through mechanical drawing process is usually carried out to afford adequate mechanical performance [64]. The mechanical drawing process aligns molecular orientations along a filament axis that imparts mechanical properties. Kelnar et al. [65] investigated the effect of the cold and hot drawing of PLA/PCL-GNP composites on the resultant mechanical performance. Melt-drawn samples at draw ratio (L/L_0) of three had a tensile modulus of ~0.97 N tex^{-1}, tenacity of ~3.12 cN tex^{-1} and elongation of 570%. The cold-drawn samples, i.e., draw ratio of 4 (CDL) and 6 (CDH), exhibited a tensile modulus of 0.92 N tex^{-1} and 1.13 N tex^{-1}, tenacity of 18.1 cN tex^{-1} and 2.35 cN tex^{-1}, and elongation of 41.3% and 88.2%, respectively. The improvement of tenacity for CDL was because of the low extent of drawing and because of a lower velocity that afforded oriented PLA fibrils formation within the sample. However, higher extent and velocity of cold drawing do not allow sufficient time for the fibril formation. Similarly, melt-drawing results in lower PLA fibrillation within the sample.

6.4. Hybridization

Hybridization is one of the conducive solutions to overcome the issues related to single-reinforced (nano)composites [66-70]. This process is often used not only to contribute to the overall properties of the resulting composite materials, but to promote the dispersion of the fillers [71-73]. The facilitation of dispersion by the secondary fillers further improves the overall properties of the resultant composites, especially mechanical performance. For instance, the inclusion of GNPS into CNT-reinforced composites reduces the overall production costs while maintaining anticipated properties [66]. Tait et al. [66] compared the preparation method to the mechanical properties of carbon filler and their hybrids. It was reported that there was no filler synergy when both fillers were incorporated into PLA. However, vapour-grown carbon fibres (VGCFs) increased the flexural strength; meanwhile GNPS strongly affected flexural modulus. For processing methods, it was difficult to reach a conclusion, but solution-casting improved overall mechanical properties at higher filler loadings, while extrusion is preferable for lower loadings and neat PLA.

Han et al. [67] prepared hybrid composites to investigate possible synergy between the filler using extrusion followed by injection moulding. In this case, GNPs were dispersed in isopropyl alcohol using sonicator, followed by the introduction of kenaf fibres as second filler. After evaporating isopropyl alcohol, kenaf fibres were melt-extruded at 180°C and a speed of 150 rpm with a residence time of 3 minutes. It was reported that the addition of GNPs into PLA had no significant influence on the flexural strength, but the inclusion of kenaf alone led to a significant increase of up to 30% loading. The synergy between GNPs and kenaf was only observed when kenaf fibre loading is 40%, but between 10-30% loading, kenaf-based composites performed better. This was attributed to GNPs coating kenaf fibres reducing their interaction with PLA and thus hindering stress transfer between PLA and fibres. In the case of flexural modulus, the presence of GNPs further improved flexural modulus when incorporated into PLA/kenaf. The incorporation of 5% GNPs was enhanced by a modulus of PLA by 25%, and 40% kenaf by ~110%, but their combination (PLA/40%kenaf/5%GNPs) led to 165%. This demonstrates the synergistic effect between the fillers when it comes to modulus. This was justified by an increase in storage modulus measured at room temperature using DMA. The increase in storage modulus reached ~19% with inclusion of only 5% GNPs, whereas inclusion of 40% kenaf fibres led to 62% increase. The inclusion of both GNPs and kenaf fibres, however, resulted to storage modulus increasing by 97%. This was ascribed to the synergistic reinforcing effect of the fillers with GNPs promoting the dispersion of the fibres.

The synergy between GNPs and carbon fibres on the 3D-printed composites was studied by Basheer and Marimuthu [62]. The authors report that the presence of GNP/CB improved the tensile strength of PLA. It was, however, found that the printed components exhibited high tensile strength when compared to the feeding composite material. The tensile strength increased from 31 MPa to ~63 MPa. Elsewhere, it was found that the hybridization of GNPs and nanohydrypatite (NHA) as reinforcement of PLA was depended on the content of GNPs [68]. The incorporation of 0.01% GNPs into the PLA/NHA resulted in decreased tensile strength, while impact strength and viscoelastic properties increased significantly when compared to neat PLA and PLA/NHA composites. A further increase in GNPs loading negatively affected impact strength and viscoelasticity of the hybrid composite.

The hybrid of clay nanoparticles and CNTs up to 5% was introduced to PLA to investigate the capability of the fillers to prevent mechanical properties loss due to exposure to prolong photo-ageing process [69]. Elastic modulus for reinforced PLA was higher than that of neat PLA due to the reinforcing effect of the fillers. The exposure of the PLA specimens to photodegradation, however, decreased with an increase in irradiation time because of the polymer chain cleavage. The introduction of the hybrid fillers preserved the tensile modulus at any filler content with the elastic modulus being constant up to 300 hours' irradiation exposure. This behaviour can be explained by the formation of the interconnected network of PLA chains with hybrid fillers. This results in protecting and shielding chain cleavage of the samples throughout investigated irradiation times. As mentioned earlier, the combination of CNTs with other fillers promotes the overall dispersion of the hybrid filler due to synergistic interaction between the fillers [71]. The Young's modulus increased with an increase in hybrid filler was observed due to good dispersion that allows stress transfer from PLA to the filler and the reinforcing effect of the filler [71]. It was reported that the optimal content was 0.5% for the hybrid filler to achieve the highest tensile modulus (i.e., 152 GPa for PLA and 1.76 GPa for hybrid composite) and tensile strength (i.e., from 35.5 for neat PLA to 43.8 MPa). The reduction in tensile properties at higher hybrid filler content was attributed to the self-aggregation of CNTs, which affects the stress between the filler and PLA negatively.

In their study, Bowen et al. [72] hybridized CNTs with cellulose nanocrystals (CNC) to afford composites with high electromagnetic interference shielding. CNC was used as dispersant for CNTs, which maintained a high tensile strength of ~46 MPa and tensile modulus of ~3.2 GPa. In their study, Petrény and co-workers [51] investigated the influence of the secondary processing method, i.e., 3D printing and injection moulding, on the mechanical properties of hybrid PLA composites (Figure 6.4a). In this case, carbon fibres were introduced into the CNTs/PLA composite to produce hybrid composites with high electrical conductivity. The authors used a twin-screw extruder to produce pellets that were found to be processable with injection moulding only, but when plasticizer (oligomeric lactic acid) was introduced, both secondary processing techniques afforded the production of the electro-conductive composites. Tensile strength and tensile modulus for CNT/PLA composites remained constant with an increase in CNT loading (see Figures 6.4b and d). However, there was a decrease in elongation-at-break with an increase in CNT content (Figure 6.4c). These are ascribed to the agglomeration of CNTs limiting their reinforcing effect rather acting as stress concentrating centres contributing to failure. The addition of carbon fibers resulted in materials with superior tensile strength and modulus. With an increase in CNT content, all hybrid samples showed a decrease in tensile strength, Young's modulus, and elongation-at-break. This was ascribed to CNTs leading to embrittlement of the resultant hybrid composites in which small defects and aggregates significantly to promote sample cracking and thus failure. Comparing 3D printed and injection moulded samples, it was found that the injection moulded samples exhibited superior mechanical properties. Comparing printing angles, i.e., 0° and 90°, it was pointed out that 0° printing led to better tensile modulus (~35 Nm kg^{-1} vs ~24 Nm kg^{-1}), modulus (~10721 Nm kg^{-1} vs ~6530 Nm kg^{-1}) and elongation (~0.47% vs ~0.37%).

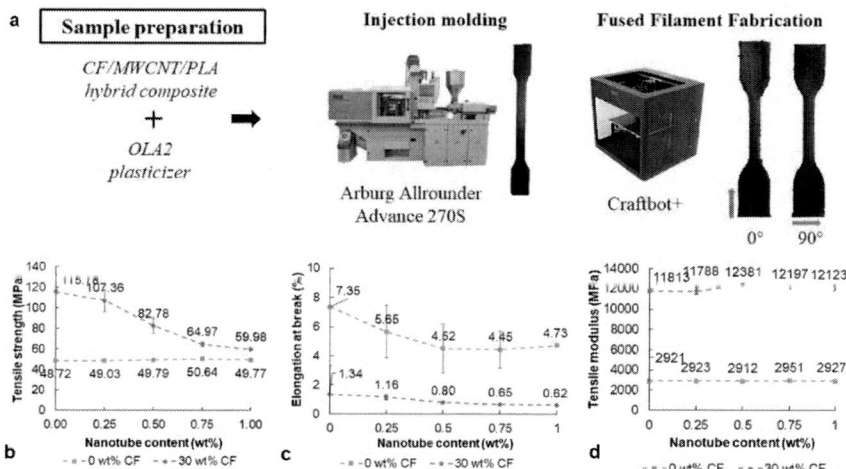

Figure 6.4. (a) Preparation of the hybrid samples; (b) tensile strength, (c) elongation-at-break and (d) Young's modulus. Reproduced with permission from Petrény et al.[51].

6.5. Conclusion

Carbon nanofillers are excellent reinforcing agents of PLA. In general, tensile modulus increases with carbon filler, regardless of the preparation method. In most cases, there is optimal carbon filler loading to achieve the desired mechanical properties. The modification of PLA using plasticizers improves toughness at the expense of tensile modulus and tensile strength. The inclusion of a carbon filler compensates for these reductions due to their reinforcing affect. The limited interaction between CBFs and PLA affect stress transfer negatively between PLA and the fillers, and thus affect composites' mechanical performance negatively. The use of carbon fillers with high-aspect ratio and their surface modification improves overall tensile properties of the resulting composite materials. The incorporation of the secondary filler improves the overall mechanical properties of the PLA/CBF composites. The optimal content of the hybrid filler is essential to avoid irreversible aggregation and achieve the desired mechanical properties. There is a need for studies on the use of hybrid fillers in 3D-printing applications in order to evaluate the resulting properties, since the printed components' properties are influenced by the feeding materials. The hybridization process facilitates the dispersion and interfacial adhesion which can significantly improve the overall properties of the 3D-printed components.

References

[1] Nair LS, Laurencin CT. Biodegradable polymers as biomaterials. *Progress in Polymer Science* 2007; 32:762-98.
[2] Mokhena T, Mochane M. Transforming fishery waste into chitin and chitin-based materials. *"Waste-to-Profit"?(WtP): Value Added Products to Generate Wealth for a Sustainable Economy* 2018:227-49.
[3] Mokhena TC, Jacobs V, Luyt AS. A review on electrospun bio-based polymers for water treatment. *Express Polym Lett* 2015; 9:839-80.
[4] Mokhena TC, Sadiku ER, Ray SS, Mochane MJ, Matabola KP, Motloung M. Flame retardancy efficacy of phytic acid: An overview. *Journal of Applied Polymer Science* 2022; 139:e52495.
[5] Mokhena TC, John MJ. Esterified cellulose nanofibres from saw dust using vegetable oil. *International Journal of Biological Macromolecules* 2020; 148:1109-17.
[6] Mokhena TC, John MJ. Cellulose nanomaterials: new generation materials for solving global issues. *Cellulose* 2020; 27:1149-94.
[7] Mokhena TC, Sadiku ER, Mochane MJ, Ray SS, John MJ, Mtibe A. Mechanical properties of cellulose nanofibril papers and their bionanocomposites: A review. *Carbohydrate Polymers* 2021; 273:118507.
[8] Darder M, Aranda P, Ruiz-Hitzky E. Bionanocomposites: A New Concept of Ecological, Bioinspired, and Functional Hybrid Materials. *Advanced Materials* 2007; 19:1309-19.
[9] Mokhena TC, Sadiku ER, Ray SS, Mochane MJ, Motaung TE. The effect of expanded graphite/clay nanoparticles on thermal, rheological, and fire-retardant properties of poly(butylene succinate). *Polymer Composites* 2021; 42:6370-82.
[10] Mokhena TC, Mochane MJ, Sadiku ER, Agboola O, John MJ. Opportunities for PLA and Its Blends in Various Applications. In: Gnanasekaran D, editor. *Green Biopolymers and their Nanocomposites.* Singapore: Springer Singapore; 2019. p. 55-81.
[11] Mochane MJ, Mokhena TC, Sadiku ER, Ray SS, Mofokeng TG. Green Polymer Composites Based on Polylactic Acid (PLA) and Fibers. In: Gnanasekaran D, editor. *Green Biopolymers and their Nanocomposites.* Singapore: Springer Singapore; 2019. p. 29-54.

- [12] Mokhena TC, Sefadi JS, Sadiku ER, John MJ, Mochane MJ, Mtibe A. Thermoplastic Processing of PLA/Cellulose Nanomaterials Composites. *Polymers* 2018;10.
- [13] Jariyavidyanont K, Yu Q, Petzold A, Thurn-Albrecht T, Glüge R, Altenbach H, et al. Young's modulus of the different crystalline phases of poly (l-lactic acid). *Journal of the Mechanical Behavior of Biomedical Materials 2023*; 137:105546.
- [14] Litauszki K, Petrény R, Haramia Z, Mészáros L. Combined effects of plasticizers and D-lactide content on the mechanical and morphological behavior of polylactic acid. *Heliyon* 2023; 9:e14674.
- [15] Dizon JRC, Espera AH, Chen Q, Advincula RC. Mechanical characterization of 3D-printed polymers. *Additive Manufacturing* 2018; 20:44-67.
- [16] Rane AV, Kanny K, Mathew A, Mohan TP, Thomas S. Comparative Analysis of Processing Techniques' Effect on the Strength of Carbon Black (N220)-Filled Poly (Lactic Acid) Composites. *Strength of Materials* 2019; 51:476-89.
- [17] Zhou C, Guo H, Li J, Huang S, Li H, Meng Y, et al. Temperature dependence of poly(lactic acid) mechanical properties. *RSC Advances* 2016; 6:113762-72.
- [18] Cocca M, Lorenzo MLD, Malinconico M, Frezza V. Influence of crystal polymorphism on mechanical and barrier properties of poly(l-lactic acid). *European Polymer Journal* 2011; 47:1073-80.
- [19] Tábi T, Ageyeva T, Kovács JG. Improving the ductility and heat deflection temperature of injection molded Poly(lactic acid) products: A comprehensive review. *Polymer Testing* 2021; 101:107282.
- [20] Puiggali J, Ikada Y, Tsuji H, Cartier L, Okihara T, Lotz B. The frustrated structure of poly(l-lactide). *Polymer* 2000; 41:8921-30.
- [21] Zhang J, Duan Y, Sato H, Tsuji H, Noda I, Yan S, et al. Crystal Modifications and Thermal Behavior of Poly(l-lactic acid) Revealed by Infrared Spectroscopy. *Macromolecules* 2005; 38:8012-21.
- [22] Cartier L, Okihara T, Ikada Y, Tsuji H, Puiggali J, Lotz B. Epitaxial crystallization and crystalline polymorphism of polylactides. *Polymer* 2000; 41:8909-19.
- [23] Yoon JT, Jeong YG, Lee SC, Min BG. Influences of poly(lactic acid)-grafted carbon nanotube on thermal, mechanical, and electrical properties of poly(lactic acid). *Polymers for Advanced Technologies* 2009; 20:631-8.
- [24] Yoon JT, Lee SC, Jeong YG. Effects of grafted chain length on mechanical and electrical properties of nanocomposites containing polylactide-grafted carbon nanotubes. *Composites Science and Technology* 2010; 70:776-82.
- [25] Chrissafis K, Paraskevopoulos KM, Jannakoudakis A, Beslikas T, Bikiaris D. Oxidized multiwalled carbon nanotubes as effective reinforcement and thermal stability agents of poly(lactic acid) ligaments. *Journal of Applied Polymer Science* 2010; 118:2712-21.
- [26] Jaseem SM, Ali NA. Antistatic packaging of carbon black on plastizers biodegradable polylactic acid nanocomposites. *Journal of Physics: Conference Series* 2019; 1279:012046.

[27] Kim H-S, Hyun Park B, Yoon J-S, Jin H-J. Thermal and electrical properties of poly(l-lactide)-graft-multiwalled carbon nanotube composites. *European Polymer Journal* 2007; 43:1729-35.
[28] Zhang L, Li Y, Wang H, Qiao Y, Chen J, Cao S. Strong and ductile poly(lactic acid) nanocomposite films reinforced with alkylated graphene nanosheets. *Chemical Engineering Journal* 2015; 264:538-46.
[29] Su Z, Guo W, Liu Y, Li Q, Wu C. Non-isothermal crystallization kinetics of poly (lactic acid)/modified carbon black composite. *Polymer Bulletin* 2009; 62:629-42.
[30] Chiu W-M, Kuo H-Y, Tsai P-A, Wu J-H. Preparation and Properties of Poly (Lactic Acid) Nanocomposites Filled with Functionalized Single-Walled Carbon Nanotubes. *Journal of Polymers and the Environment* 2013; 21:350-8.
[31] Wang L-N, Guo Wang P-Y, Wei J-C. Graphene Oxide-Graft-Poly(l-lactide)/Poly(l-lactide) Nanocomposites: Mechanical and Thermal Properties. *Polymers* 2017.
[32] Pramoda KP, Koh CB, Hazrat H, He CB. Performance enhancement of polylactide by nanoblending with POSS and graphene oxide. *Polymer Composites* 2014; 35:118-26.
[33] Chiesa E, Dorati R, Pisani S, Bruni G, Rizzi LG, Conti B, et al. Graphene Nanoplatelets for the Development of Reinforced PLA–PCL *Electrospun Fibers as the Next-Generation of Biomedical Mats. Polymers* 2020.
[34] Zhao X, Luo J, Fang C, Xiong J. Investigation of polylactide/poly(ε-caprolactone)/multi-walled carbon nanotubes electrospun nanofibers with surface texture. *RSC Advances* 2015; 5:99179-87.
[35] Wu D, Samanta A, Srivastava RK, Hakkarainen M. Nano-Graphene Oxide Functionalized Bioactive Poly(lactic acid) and Poly(ε-caprolactone) Nanofibrous Scaffolds. *Materials* 2018.
[36] Wang X, Zhuang Y, Dong L. Study of carbon black-filled poly(butylene succinate)/polylactide blend. *Journal of Applied Polymer Science* 2012; 126:1876-84.
[37] Al Zahmi S, Alhammadi S, ElHassan A, Ahmed W. Carbon Fiber/PLA Recycled Composite. *Polymers* 2022.
[38] Guo J, Tsou C-H, Yu Y, Wu C-S, Zhang X, Chen Z, et al. Conductivity and mechanical properties of carbon black-reinforced poly(lactic acid) (PLA/CB) composites. *Iranian Polymer Journal* 2021; 30:1251-62.
[39] Delgado PA, Brutman JP, Masica K, Molde J, Wood B, Hillmyer MA. High surface area carbon black (BP-2000) as a reinforcing agent for poly[(−)-lactide]. *Journal of Applied Polymer Science* 2016;133.
[40] Chieng BW, Ibrahim NA, Yunus WM, Hussein MZ. Poly(lactic acid)/Poly(ethylene glycol) Polymer Nanocomposites: Effects of Graphene Nanoplatelets. *Polymers* 2014. p. 93-104.
[41] Chieng BW, Ibrahim NA, Wan Yunus WMZ, Hussein MZ, Loo YY. Effect of graphene nanoplatelets as nanofiller in plasticized poly(lactic acid) nanocomposites. *Journal of Thermal Analysis and Calorimetry* 2014; 118:1551-9.
[42] Chieng BW, Ibrahim NA, Wan Yunus WMZ, Hussein MZ, Silverajah VSG. Graphene Nanoplatelets as Novel Reinforcement Filler in Poly(lactic

acid)/Epoxidized Palm Oil Green Nanocomposites: Mechanical Properties. *International Journal of Molecular Sciences* 2012; 13:10920-34.

[43] Wang N, Zhang X, Ma X, Fang J. Influence of carbon black on the properties of plasticized poly(lactic acid) composites. *Polymer Degradation and Stability* 2008; 93:1044-52.

[44] Yu J, Wang N, Ma X. Fabrication and Characterization of Poly(lactic acid)/Acetyl Tributyl Citrate/Carbon Black as Conductive Polymer Composites. *Biomacromolecules* 2008; 9:1050-7.

[45] Chieng BW, Ibrahim NA, Yunus WM, Hussein MZ, Then YY, Loo YY. Effects of Graphene Nanoplatelets and Reduced Graphene Oxide on Poly(lactic acid) and Plasticized Poly(lactic acid): A Comparative Study. *Polymers* 2014. p. 2232-46.

[46] Batakliev T, Petrova-Doycheva I, Angelov V, Georgiev V, Ivanov E, Kotsilkova R, et al. Effects of Graphene Nanoplatelets and Multiwall Carbon Nanotubes on the Structure and Mechanical Properties of Poly(lactic acid) Composites: A Comparative Study. *Applied Sciences* 2019.

[47] Gao Y, Picot OT, Bilotti E, Peijs T. Influence of filler size on the properties of poly(lactic acid) (PLA)/graphene nanoplatelet (GNP) nanocomposites. *European Polymer Journal* 2017; 86:117-31.

[48] Wang Y, Mei Y, Wang Q, Wei W, Huang F, Li Y, et al. Improved fracture toughness and ductility of PLA composites by incorporating a small amount of surface-modified helical carbon nanotubes. *Composites Part B: Engineering* 2019; 162:54-61.

[49] Wu G, Liu S, Wu X, Ding X. Core-shell structure of carbon nanotube nanocapsules reinforced poly(lactic acid) composites. *Journal of Applied Polymer Science* 2017;134.

[50] Marconi S, Alaimo G, Mauri V, Torre M, Auricchio F. Impact of graphene reinforcement on mechanical properties of PLA 3D printed materials. *2017 IEEE MTT-S International Microwave Workshop Series on Advanced Materials and Processes for RF and THz Applications* (IMWS-AMP)2017. p. 1-3.

[51] Petrény R, Tóth C, Horváth A, Mészáros L. Development of electrically conductive hybrid composites with a poly(lactic acid) matrix, with enhanced toughness for injection molding, and material extrusion-based additive manufacturing. *Heliyon* 2022; 8:e10287.

[52] Kotlinski J. Mechanical properties of commercial rapid prototyping materials. *Rapid Prototyping Journal* 2014; 20:499-510.

[53] Chacón JM, Caminero MA, García-Plaza E, Núñez PJ. Additive manufacturing of PLA structures using fused deposition modelling: Effect of process parameters on mechanical properties and their optimal selection. *Materials & Design* 2017; 124:143-57.

[54] Aloyaydi B, Sivasankaran S, Mustafa A. Investigation of infill-patterns on mechanical response of 3D printed poly-lactic-acid. *Polymer Testing* 2020; 87:106557.

[55] Ayatollahi MR, Nabavi-Kivi A, Bahrami B, Yazid Yahya M, Khosravani MR. The influence of in-plane raster angle on tensile and fracture strengths of 3D-printed PLA specimens. *Engineering Fracture Mechanics* 2020; 237:107225.

[56] Cicero S, Martínez-Mata V, Castanon-Jano L, Alonso-Estebanez A, Arroyo B. Analysis of notch effect in the fracture behaviour of additively manufactured PLA and graphene reinforced PLA. *Theoretical and Applied Fracture Mechanics* 2021; 114:103032.

[57] Lanzotti A, Grasso M, Staiano G, Martorelli M. The impact of process parameters on mechanical properties of parts fabricated in PLA with an open-source 3-D printer. *Rapid Prototyping Journal* 2015; 21:604-17.

[58] Caminero MÁ, Chacón JM, García-Plaza E, Núñez PJ, Reverte JM, Becar JP. Additive Manufacturing of PLA-Based Composites Using Fused Filament Fabrication: Effect of Graphene Nanoplatelet Reinforcement on Mechanical Properties, Dimensional Accuracy and Texture. *Polymers* 2019.

[59] Bortoli LSD, Farias Rd, Mezalira DZ, Schabbach LM, Fredel MC. Functionalized carbon nanotubes for 3D-printed PLA-nanocomposites: Effects on thermal and mechanical properties. *Materials Today Communications* 2022; 31:103402.

[60] Zhou X, Deng J, Fang C, Lei W, Song Y, Zhang Z, et al. Additive manufacturing of CNTs/PLA composites and the correlation between microstructure and functional properties. *Journal of Materials Science & Technology* 2021; 60:27-34.

[61] Zhou Y, Lei L, Yang B, Li J, Ren J. Preparation and characterization of polylactic acid (PLA) carbon nanotube nanocomposites. *Polymer Testing* 2018; 68:34-8.

[62] Mohammed Basheer EP, Marimuthu K. Carbon fibre-graphene composite polylactic acid (PLA) material for COVID shield frame. *Materialwissenschaft und Werkstofftechnik* 2022; 53:119-27.

[63] Patanwala HS, Hong D, Vora SR, Bognet B, Ma AWK. The microstructure and mechanical properties of 3D printed carbon nanotube-polylactic acid composites. *Polymer Composites* 2018; 39:E1060-E71.

[64] Rawal A, Mukhopadhyay S. 4 - Melt spinning of synthetic polymeric filaments. In: Zhang D, editor. *Advances in Filament Yarn Spinning of Textiles and Polymers:* Woodhead Publishing; 2014. p. 75-99.

[65] Kelnar I, Zhigunov A, Kaprálková L, Fortelný I, Dybal J, Kratochvíl J, et al. Facile preparation of biocompatible poly (lactic acid)-reinforced poly(ε-caprolactone) fibers via graphite nanoplatelets -aided melt spinning. *Journal of the Mechanical Behavior of Biomedical Materials* 2018; 84:108-15.

[66] Tait M, Pegoretti A, Dorigato A, Kalaitzidou K. The effect of filler type and content and the manufacturing process on the performance of multifunctional carbon/poly-lactide composites. *Carbon* 2011; 49:4280-90.

[67] Han SO, Karevan M, Bhuiyan MA, Park JH, Kalaitzidou K. Effect of exfoliated graphite nanoplatelets on the mechanical and viscoelastic properties of poly(lactic acid) biocomposites reinforced with kenaf fibers. *Journal of Materials Science* 2012; 47:3535-43.

[68] Michael FM, Khalid M, Chantara Thevy R, Raju G, Shahabuddin S, Walvekar R, et al. Graphene/Nanohydroxyapatite hybrid reinforced polylactic acid nanocomposite for load-bearing applications. *Polymer-Plastics Technology and Materials* 2022; 61:803-15.

[69] Gorrasi G, Milone C, Piperopoulos E, Lanza M, Sorrentino A. Hybrid clay mineral-carbon nanotube-PLA nanocomposite films. Preparation and photodegradation effect on their mechanical, thermal and electrical properties. *Applied Clay Science* 2013; 71:49-54.

[70] Santangelo S, Gorrasi G, Di Lieto R, De Pasquale S, Patimo G, Piperopoulos E, et al. Polylactide and carbon nanotubes/smectite-clay nanocomposites: Preparation, characterization, sorptive and electrical properties. *Applied Clay Science* 2011;5 3:188-94.

[71] Sanusi OM, Benelfellah A, Papadopoulos L, Terzopoulou Z, Malletzidou L, Vasileiadis IG, et al. Influence of montmorillonite/carbon nanotube hybrid nanofillers on the properties of poly(lactic acid). *Applied Clay Science* 2021; 201:105925.

[72] Yu B, Zhao Z, Fu S, Meng L, Liu Y, Chen F, et al. Fabrication of PLA/CNC/CNT conductive composites for high electromagnetic interference shielding based on Pickering emulsions method. *Composites Part A: Applied Science and Manufacturing* 2019; 125:105558.

[73] Petrény, Mészáros. Multi-scale hybrid composites are making their way. *Express Polym Lett* 2022; 16:1228-.

Chapter 7

Applications of PLA/Carbon Based Fillers Composites

Abstract

PLA can be produced from different natural resources (i.e., maize, sugarcane, sugar beet and cassava), which results in materials with unique properties, such as sustainability and renewability. Depending on the synthesis route and the race mixture ratio, different grades of PLA can be obtained that qualify its use in industrial and commodity applications. It has been widely used in agriculture, packaging, electronics, medicine, biotechnology, building and home furnishing. The current chapter reports on the application of PLA and PLA/carbon-based fillers.

Keywords: packaging, health, sensing, mask, oil-water separation

7.1. Introduction

In the current society, there is a need for the fabrication of polymeric materials that meet the societal needs. The major utilization of the polymers and their composites includes applications in sports, airplane, medicinal, gadgets, home appliances and automotives [1]. The ease of manufacturing of polymers and their composites is one of the reasons for their expanding applications. There is an increase in environmental pollution and generation of waste from the non-biodegradable polymer-based materials [2]. The fact that non-biodegradable plastics take years to degrade means that they have a long effect on humans and the environment at large. In most cases, their decomposition include fire, which releases noxious vapor. The well-known non-biodegradable polymers products include rayon, nylon, poly-vinyl chloride (PVC), Lexan and polyester. The disposing of non-biodegradable plastics requires approximately 400 years to degrade. Figure 7.1 illustrates the degradation time for various plastics.

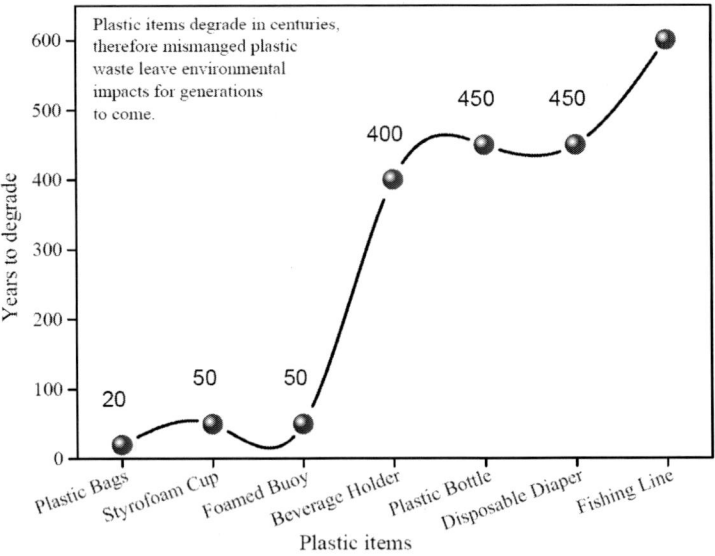

Figure 7.1. The degradation time of selective plastics [3].

Based on the above time frame for degradation of non-biodegradable polymers, there is a huge need for applications of biodegradable polymers in various sectors for a sustainable future. Polylactic acid is one of the biodegradable polymers that is favoured for advanced applications because it is environmentally friendly, biocompatible, and biodegradable. As such it would reduce the environmental problems associated with non-biodegradable plastic wastes that are produced daily. It is one of the most explored biopolymers due to its biocompatibility, biodegradability, compostability, and nontoxicity [4-5]. In order to broaden PLA applications, improvements of physical properties, i.e., mechanical, thermal stability, and electrical conductivity are required. The inclusion of different fillers is the most commonly used approach to fine-tuned physical properties depending on the filler-type and loading [4]. Carbon nanofillers received a great deal of attention as filler of choice to improve the properties of PLA [6-8]. A number of methods have been developed for syntheses of carbon nanofillers that have been used to prepare PLA (nano)composites [8-9]. These methods result in fillers with desirable functionalities to promote their interaction and dispersion within the PLA matrix. The resulting multifunctional composites are extensively applied in smart sensors. These composites are reported to be applicable in packaging, tissue engineering and drug delivery because of their features, including good barrier properties, excellent mechanical properties and surface tailor-ability to afford bio-conjugation with desirable compounds. The current chapter provides an in-depth discussion on the applications of PLA.

7.2. Applications of CBF-PLA Composites

7.2.1. Packaging

One of the prerequisites for food packaging is flexibility and good barrier performance to eliminate permeation gases that cause the spoiling of foods [10-12]. For instance, the permeation of oxygen can result in lipid oxidation. The incorporation of fillers can prolong the shelf-life of food due to tortuous pathways created by the presence of impermeable fillers [10]. Four different processes are involved in the gas permeation through the specimen: sorption of gas molecules onto the surface, dissolution of gases into the specimen, diffusion of gas through the specimen, and desorption of the gas on the opposite side of the specimen [13]. By reducing the gas-dissolving coefficient, allowing for longer and more tortuous gas diffusion, and increasing the gas-permeation lag time and decreasing the gas-diffusion coefficient, gas permeation can be reduced [13]. The permeation of gas in thin packaging films is schematically presented in Figure 7.2.

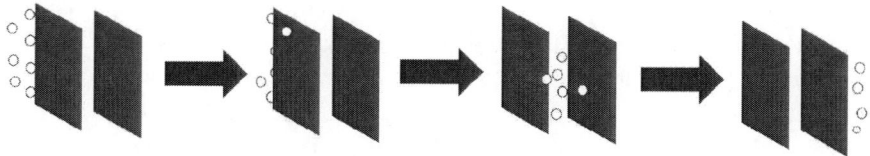

Figure 7.2. Schematic presentation of gas permeation process in thin films. Reproduced from [14]. In this case, the gas-permeation coefficient (P, mol Pa^{-1} m^{-1} s^{-1}) can be calculated according to Equation 7.1 [14, 15].

$$P = D \times S \qquad 7.1$$

where diffusion coefficient is presented by D (m^{2}s^{-1}) and S (mol m^{-3} Pa^{-1}) presents the solubility coefficient of the gas on the surface of the specimen (see Figure 7.2). The coefficient D of the polymer is given by Equation 7.2 [14, 15]:

$$D = \frac{L^2}{6t_L} \qquad 7.2$$

where L is the specimen thickness and t_L is the gas-permeation lag time. The gas permeation can be influenced by the size of the filler, as demonstrated by Jiang et al. [14]. The authors introduced large-scale graphene oxide (GO-L) and small-scale graphene oxide (GO-S) into PLA using solution flocculation to afford their uniform dispersion. The oxygen permeability (PO$_2$) reduced from 3.30×10^{-14} cm$^3 \cdot$cm$\cdot$cm$^{-2} \cdot$s$^{-1} \cdot$Pa^{-1} to 3.11×10^{-14} cm$^3 \cdot$cm$\cdot$cm$^{-2} \cdot$s$^{-1} \cdot$Pa^{-1} and 3.01×10^{-14} cm$^3 \cdot$cm$\cdot$cm$^{-2} \cdot$s$^{-1} \cdot$Pa^{-1} for GO-S and GO-L based composites, respectively (Figure 7.3). The decrease of gas permeability by 5.8% and 8.8% for GO-S and GO-L-based composites demonstrates that larger particles are preferable to decreasing gas permeability and improve overall barrier properties performance. A recent study by Cruz et al. [16] demonstrates that the modification of graphene oxide can improve the affinity of GO to PLA, which leads to improved overall properties of the resultant composites. Graphene oxide was functionalized with two types of alkylamines, i.e., decylamine (DA) and octadecylamine (ODA) synthesized at two different temperatures (25°C and 80°C). Water vapour and oxygen permeability (P_{O2}) decreased with an increase in filler content. The modification of GO with ODA at 80°C exhibited superior P_{O2} and water vapour permeability with a ~30% reduction for P_{O2} barrier, with only 0.7%GO loading, whereas a 50% reduction in water vapour recorded at 0.2%GO loading.

In case of flexibility, different polymers and plasticizers are usually introduced into PLA-based materials to enhance the toughness and ductility. Pinto et al. [11] report that the presence of graphene oxide and graphene nanoplatelets decreases oxygen and nitrogen permeability three- and four-fold when compared to neat PLA. This is attributed to the barrier effect created by the impermeable fillers. Good dispersibility of the fillers resulted in high transparency, making these films suitable for packaging. The plasticized PLA with thymol elevated the toughness of the resulting blend [17]. The incorporation of carbon-black improved the flexibility as well as the antistatic properties of PLA/thymol blends. It was reported that elongation-at-break increased with CB content, thus demonstrating the potential of these composites for antistatic packaging purposes. Highly flexible PLA was achieved by introducing octadecylamine (ODA) functionalized graphene oxide (GO-ODA). The (nano)composites exhibited a 34% increase in tensile strength, 44% in tensile modulus, and a 300%

increase in toughness, with only 0.4% loading [18]. This results from strong interfacial adhesion, which contributed to the homogeneous dispersion of the fillers, and thus improved thermal stability and enhanced crystallization behaviour of PLA, indicating that composites can be used for smart packaging applications.

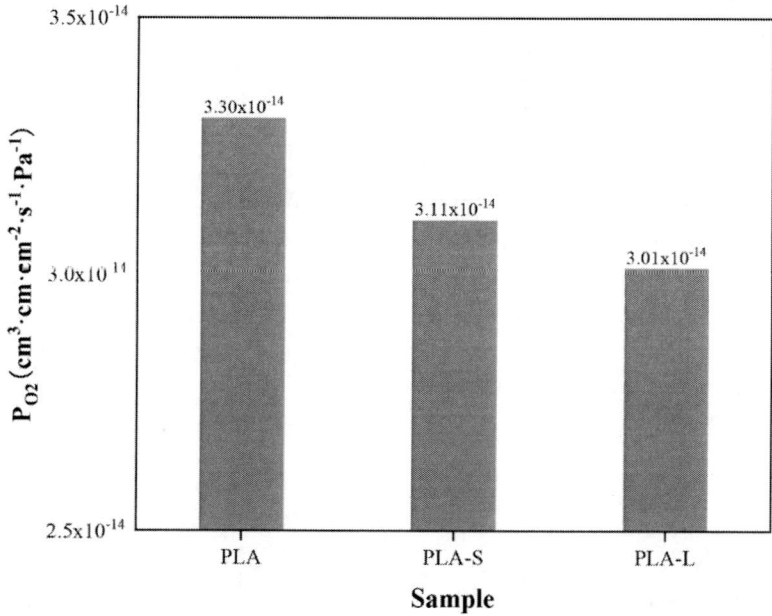

Figure 7.3. Oxygen Permeation for neat PLA and nanocomposites (PLA-S and PLA-L is smaller GO and larger GO fillers, respectively. Reproduced with permission from Jiang, et al.[14].

Hybridization of carbon (nano)fillers for packaging purposes was reported in literature [10]. In this case, CNT and GO hybrid filler was introduced into PLA to fabricate thin films using solution casting. It was found that the hybrid fillers were homogeneously dispersed, which led to excellent blockage of visible light, more than single-reinforced GO and CNT-based composites. The transmittance at 250 nm and 300 nm reduced to 40% and 50% with the addition of 0.4%GO-CNT loading. The oxygen transmission rate (OTR) decreased from 729 mL m^{-2}day^{-1} to 244 mL m^{-2}day^{-1} because of the tortuous diffusion route created by impermeable hybrid filler. These results demonstrate that these composites can be used readily in different packaging systems. Hybrid filler of graphene oxide loaded with zinc oxide nanoparticles (ZnONPs) was studied as suitable filler to enhance the anti-UV and antimicrobial properties of PLA [19]. The resulting films exhibited good anti-UV with excellent transmittance under visible light. The increase in hybrid composite resulted in an increase in anti-UV property while good transmittance was retained. The antibacterial efficacy in the dark was 52.3% for *E. coli* and 83% for *S. aureus*; meanwhile 99.2 and 97.6% for *S. aureus* and *E. coli* under light illumination is achieved with only 0.2% GO-ZnONPs loading.

7.2.2. Healthcare Applications

The exploitation of PLA-based materials in healthcare is because of its biocompatibility. It can easily be broken down in the human body into non-toxic and non-hazardous lactic acid that can be metabolized by the body. The addition of the carbon nanofillers into PLA can broaden its use in healthcare for tissue engineering, bone engineering, sutures, nappies, hospital gowns, drug-delivery carriers and many others.

7.2.2.1. Tissue Engineering

The electrospinning technique produces fibres with diameters ranging between nano- and micro-scale [20-22]. The resulting morphology with high porosity closely mimics the natural extracellular matrix (ECM) to afford their application in tissue engineering [20, 22]. The reinforcement of PLA/PEG with graphene oxide (GO) for tissue engineering was reported by Zhang et al. [20]. From the assessment of electro-spun nanofibers in tissue engineering using NIH 3T3 cells, it was observed that all nanofibers were covered with living cells, with few dead cells being depicted. The cells spreading on the surface of the fibres were associated with scaffolds, providing a conducive environment for cell viability and proliferation. It was found that PLA/2%GO-g-PEG provided more cell viability and proliferation, indicating that this scaffold is suitable for cell growth. The mechanical properties of the resulting nanocomposites qualify these fibres in tissue engineering applications. In most cases, the surface wettability is a crucial aspect for cell attachment and growth [21]. GO has hydroxyl and carboxylic groups which can impart hydrophilicity of PLA for tissue engineering applications [21]. The increase in hydrophilicity was observed to increase in GO content within electro-spun PLA composites (see Figure 7.4a) [21]. Using osteoblast MG63 cell lines, it was found that the presence of 1% and 2.5%GO within electro-spun PLA fibre exhibited good cell viability higher than 120% demonstrating good cell compatibility (Figure 7.4b). The cells homogeneously adhered to the fibre surface (Figure 7.4c&d).

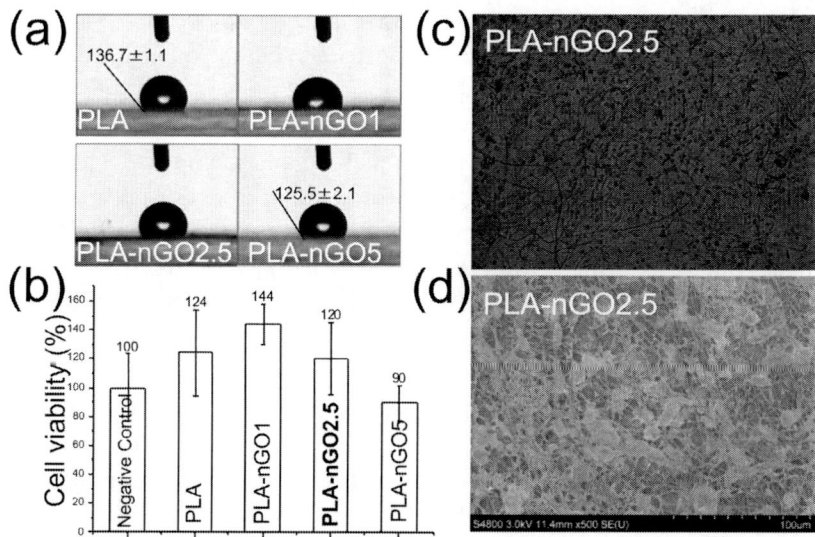

Figure 7.4. (a) Water contact angle for neat PLA, 1%GO/PLA (PLA/NGO1), 2.5%GO/PLA (PLA-nGO2.5) and 5%GO/PLA (PLA-nGO5), (b) cell viability (c) optical microscopy and (d) SEM images of 2.5%GO/PLA fibres (PLA-nGO2.5) after 4 days cell culturing. Reproduced from Wu et al.[21].

Zhao et al. [23] incorporated CNTs into electro-spun PLA/PCL blends for tissue engineering applications. CNTs were used as compatibilizers between immiscible an PLA/PCL blend to afford parallel-line, surface-textured electro-spun nanofibers. The presence of CNTs improves the mechanical performance of the resultant electro-spun composites. It was found that the composite exhibited no toxic effect of the proliferation of the fibroblasts cells. The cells were found to spread evenly on the composite scaffold; hence they can be employed in tissue engineering.

7.2.2.2. Mask

Several nations had to establish new laws to combat the COVID-19 outbreak [5]. This included wearing masks in public as one of the most affordable methods to combat the outbreak, especially for developing countries. These masks are usually composed of three to four layers, i.e., nonwoven outer layer, melt-spun filter layer, support layer and inner nonwoven layer [5]. All these layers are made up of melt-spun synthetic polymers, and thus the use of PLA/CBFs melt-spun fibres can be used as a suitable replacement due to its melt-spinnability using conventional techniques [24]. The use of PLA-based systems is essential to overcome issues related to plastic waste resulting from classic petroleum-based polymers. On the other hand, the 3D-printability of PLA/CBFs can be exploited to fabricate a complex structure for healthcare applications [7].

Basheer and Marimuthu [7] prepared filaments composed of a hybrid carbon fibres and graphene nanosheets for printing a face-mask frame for a COVID shield. The printed components displayed superior tensile strength than the filaments which afforded their application for COVID-19 shield frams. In addition, the biodegradability of PLA and the lightweightedness of the resulting product made it applicable for short-term applications. The incorporation of modified CNTs into PLA for 3D printing resulted in printed components with a high storage modulus at 37°C, which affords the printed components to be used in human implants [25]. The tensile properties of the 90/0-angle printed components were significantly improved when compared to neat PLA and PLA/unmodified CNT-based composites. These results demonstrate that the modification of CNTs is essential to enhance the dispersion of the nanofillers and their interaction with host polymeric materials in order to broaden their applications.

7.2.2.3. Medical Sutures

PLA has been employed to manufacture medical sutures for different types of surgery circumcision, colostomy, facial, etc. because of its attractive attributes, *viz.* non-toxicity, biocompatibility, biodegradability and high mechanical performance [5]. During wound healing, PLA decomposes into CO_2 and H_2O; hence avoiding the pain associated with suture removal. The rate of PLA degradation is an important parameter when it comes to wound healing. Some wounds need a medical suture to last for a certain period longer than the fixed degradation period of PLA. PLA blending with plasticizers and polymers has been studied as a reliable route to control the degradation rate of PLA. In most cases, the presence of these components rather increases the degradation rate of PLA. Therefore, the incorporation of carbon nanofillers offers an opportunity to extend and control the degradation rate of PLA [24]. Liu et al. [24] prepared sutures composed of CNT and PLA using a melt-spinning process. The degradation of the sutures was extended from 49 weeks to a maximum of 73 weeks with the addition of 1wt% CNTs (Figure 7.5). The inclusion of 0.5 and 2wt% CNTs extended the degradation period of the sutures from 45 weeks to 65 weeks and 63 weeks, respectively. This demonstrates that the degradation of the sutures can be prolonged by varying the content pf the carbon nanofillers within the system.

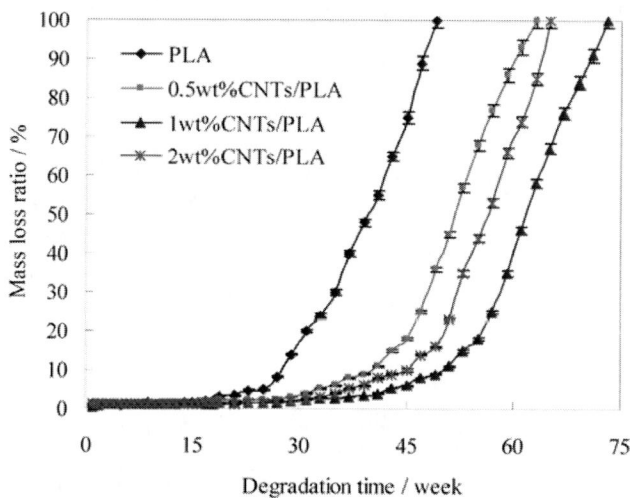

Figure 7.5. Mass loss versus degradation period. Reprinted from Liu et al. [24]. Open Access.

7.2.3. Sensing Applications

The presence of carbon nanofillers in PLA improves its electrical conductivity, which can be exploited for sensing applications [26-28]. Such composites are capable of sensing variety of stimuli, such as temperature, gases, vapour, pH, mechanical stress and strain, and liquids [29]. The sensing efficiency relies on the deformation of the carbon nanofillers percolation network from external stimuli that results in a change in electrical conductivity. Mai et al. [26] prepared PLA/CNT (nano)composites using an internal mixer, followed by a compression mould for in situ degradation monitoring. Depending on the content of the filler it was possible to follow the degradation level of the (nano)composites. All concentration close to the calculated percolation threshold exhibited four orders of magnitude of sensitivity changes with biodegradation. This improvement in sensitivity showed that this system may be utilized for monitoring degradation to offer an insight into the level of degradation essential for structural safety. The high sensitivity resulted from partial elimination of the amorphous areas from degradation of PLA.

Melt-spun PLA/CNT composites were investigated as a suitable, solvent, vapour-sensing material [27]. The sensitivity was measured by following change in electrical resistance (R_{rel}) using Equation 7.3:

$$R_{rel} = \frac{R - R_i}{R_i} \qquad 7.3$$

where R is the resistance at time t and Ri is the initial resistance. The relative resistance changes after immersion for 10 minutes decreased linearly due to the degree of network densities within the fibres. It is reported that the solvents can be distinguished with reproducible and specific relative resistance change of the fibres containing 2%CNTs. Water exhibited the maximum relative resistance of 0.01 after 10 minutes, whereas 0.74, 0.12 and 0.86 for ethanol, n-hexane and methanol, respectively were attained. The specimens were capable of distinguishing a mixture of ethanol and water. These results were even better than the ones obtained for compression-moulded samples, but the issue was the fibres retaining their initial resistance values during drying. It was indicated that these fibres could rather be used for leakage detection. Elsewhere, thin films of PLA/CNTs vapour-sensing behaviour was found to be dependent on the CNTs loadings [28]. The thin films were produced using an extrusion method followed by melt-compounding. Thin films with low loadings (0.5-0.75%) showed larger relative resistance change with high signal noises. Since the preparation method has a significant impact on the properties of the resulting PLA/CBFs composites, Kumar et al. [30] prepared composites by thorough solution mixing in order to prepare transducers via a spray layer-by-layer technique. It was found that crystallinity and filler content are crucial to achieve the desired sensing properties. The sensor had high selectivity towards chloroform, followed by methanol, with 2% as percolation threshold concentration. The increase in crystallinity decreased selectivity of the sensor.

Electrochemical analysis is one of the reliable practices for the detection of different pollutants [31, 32]. This involves the surface treatment of the electrode to achieve the detection of the desired pollutant. In most cases, conductive polymers are commonly used as electrode materials. In recent years, biopolymers (i.e., PLA, cellulose, chitosan, etc.) with their unique features, such as acceptable strength, ease casting, biocompatibility, sustainability, and biodegradability received a great deal of attention as replacement for conventional conductive polymers. Chakraborty et al. [31] studied the use of PLA/exfoliated graphene (GR) for copper ion detection. The composites were prepared using solution casting to afford thin films. These films were decorated with albumin, Ethylenediaminetetraacetic (EDTA) and silk nanocrystal (SNC) to promote the sensitivity and specificity. It was found that PLA/GR composite had good sensitivity of $0.36\mu A$ ppm^{-1} for copper (Cu) ion detection in the concentration range of 5-100 ppm. The decoration of thin films were found to enhance the selectivity without affecting the limit of detection (LOD) and sensitivity.

7.2.4. Oil-Water Separation

Water pollution is one of the most critical issues that require attention from industries and research communities [33-35]. The use of renewable and sustainability materials is essential from economic and ecological viewpoints. This is because these materials are more environmentally friendly and do not produce dangerous by products when they are disposed of after their lifespan. PLA is one of the sustainable materials with huge potential in wastewater treatment [5, 36]. One of the materials usually used for oil-water separation is aerogels [37]. These materials are highly porous with excellent hydrophobicity to afford high oil-removal efficiency. PLA-based materials have been used to prepare aerogels for oil-water separation [5, 38, 39]. It was found that PLA-based aerogels have poor oil removal efficiency and limited selectivity [38]. The incorporation of the nanofillers were reported to improve the overall performance of PLA. Carbon-based materials are often used as aerogels for various applications, but their processability has been one of major concerns [40-43]. PLA can be used to facilitate the processability of carbon materials; meanwhile carbon materials can complement PLA when it comes to oil removal efficiency. Some reports indicate that the 3D-printability of PLA can be exploited to manufacture a porous membrane structure that can be decorated with other materials to afford super-hydrophobic membranes [44]. Such membranes exhibit high oil separation efficacy >95%. This could be employed to prepare PLA/CBFs composites with tailorable surface [37]. The latter can be explored to achieve desired roughness and thus hydrophobicity to enhance oil separation efficiency.

7.3. Conclusion

Polylactic acid (PLA) is a biopolymer with good properties for utilization in packaging applications. It is realized that the modification of PLA further widens the application of the PLA. For example, in the case of flexibility, different polymers and plasticizers are usually introduced into PLA-based materials to enhance the toughness and ductility. It became clear that the addition of carbon-based fillers expands the applications of the PLA. For example, the incorporation of carbon nanofillers improves electrical conductivity of PLA, which in turn expands the applications of PLA to sensing applications. In some cases, carbon-based fillers such as graphene oxide (GO) are incorporated into the PLA matrix for specific applications such as tissue engineering. In summary, PLA and PLA-carbon-based fillers have been utilized in medical, sensing and packaging applications, etc.

References

[1] Patel VK, Kant R, Chauhan PS, Bhattacharya S. Introduction to applications of polymers and polymer composites. In *Trends in applications of polymers and polymer composites,* edited by Patel VK, Kant R, Chauhan PS, Bhattacharya. Melville, New York: AIP Publishing; 2022. pp. 1-1–1-6.

[2] Dussud C, Hudec C, George M, Fabre P, Higgs P, Bruzaud S, Delort A-M, Eyheraguibel B, Meistrtzheim A-L, Jacquin J, Cheng J, Callac N, Odobel C, Rabouille S, Ghiglione J-F. Colonization of non-biodegradable and biodegradable plastics by marine microorganism. *Frontiers in Microbiology* 2018; 9:1571 (1-13).

[3] Maitlo G, Ali I, Maitlo HA, Ali S, Unar IN, Ahmad MB, Bhutto DK, Karmani RK, Naich SUR, Sajjad RU, Ali S, Afridi MN. Plastic waste recycling, applications, and future prospects for a sustainable environment. *Sustainability* 2022; 14:11637 (1-27).

[4] Mokhena TC, Sefadi JS, Sadiku ER, John MJ, Mochane MJ, Mtibe A. Thermoplastic Processing of PLA/Cellulose Nanomaterials Composites. *Polymers* 2018;10.

[5] Li X, Lin Y, Liu M, Meng L, Li C. A review of research and application of polylactic acid composites. *Journal of Applied Polymer Science* 2023; 140: e53477.

[6] Moon S-I, Jin F, Lee C-j, Tsutsumi S, Hyon S-H. Novel Carbon Nanotube/Poly(L-lactic acid) Nanocomposites; Their Modulus, Thermal Stability, and Electrical Conductivity. *Macromolecular Symposia* 2005; 224:287-96.

[7] Mohammed Basheer EP, Marimuthu K. Carbon fibre-graphene composite polylactic acid (PLA) material for COVID shield frame. *Materialwissenschaft und Werkstofftechnik* 2022; 53:119-27.

[8] Gonçalves C, Gonçalves IC, Magalhães FD, Pinto AM. Poly(lactic acid) Composites Containing Carbon-Based Nanomaterials: A Review. *Polymers* 2017.
[9] Mokhena TC, John MJ, Sibeko MA, Agbakoba VC, Mochane MJ, Mtibe A, et al. Nanomaterials: Types, Synthesis and Characterization. In: Srivastava M, Srivastava N, Mishra PK, Gupta VK, editors. *Nanomaterials in Biofuels Research.* Singapore: Springer Singapore; 2020. p. 115-41.
[10] Kim Y, Kim JS, Lee S-Y, Mahajan RL, Kim Y-T. Exploration of hybrid nanocarbon composite with polylactic acid for packaging applications. *International Journal of Biological Macromolecules* 2020; 144:135-42.
[11] Pinto AM, Cabral J, Tanaka DAP, Mendes AM, Magalhães FD. Effect of incorporation of graphene oxide and graphene nanoplatelets on mechanical and gas permeability properties of poly(lactic acid) films. *Polymer International* 2013; 62:33-40.
[12] Wu L-L, Wang J-j, He X, Zhang T, Sun H. Using Graphene Oxide to Enhance the Barrier Properties of Poly(lactic acid) Film. *Packaging Technology and Science* 2014; 27:693-700.
[13] Compton OC, Kim S, Pierre C, Torkelson JM, Nguyen ST. Crumpled Graphene Nanosheets as Highly Effective Barrier Property Enhancers. *Advanced Materials* 2010; 22:4759-63.
[14] Jiang W, Chen D, Xie Z, Zhang Y, Hu B, Kang J, et al. Exploring the Size Effect of Graphene Oxide on Crystallization Kinetics and Barrier Properties of Poly(lactic acid). *ACS Omega* 2022; 7:37315-27.
[15] Choudalakis G, Gotsis AD. Permeability of polymer/clay nanocomposites: A review. *European Polymer Journal* 2009; 45:967-84.
[16] Cruz R, Nisar M, Palza H, Yazdani-Pedram M, Aguilar-Bolados H, Quijada R. Development of bio-degradable nanocomposites based on PLA and functionalized graphene oxide. *Polymer Testing* 2023; 124:108066.
[17] Jaseem SM, Ali NA. Antistatic packaging of carbon black on plastizers biodegradable polylactic acid nanocomposites. *Journal of Physics: Conference Series* 2019; 1279:012046.
[18] Zhang L, Li Y, Wang H, Qiao Y, Chen J, Cao S. Strong and ductile poly(lactic acid) nanocomposite films reinforced with alkylated graphene nanosheets. *Chemical Engineering Journal* 2015; 264:538-46.
[19] Huang Y, Wang T, Zhao X, Wang X, Zhou L, Yang Y, et al. Poly(lactic acid)/graphene oxide–ZnO nanocomposite films with good mechanical, dynamic mechanical, anti-UV and antibacterial properties. *Journal of Chemical Technology & Biotechnology* 2015; 90:1677-84.
[20] Zhang C, Wang L, Zhai T, Wang X, Dan Y, Turng L-S. The surface grafting of graphene oxide with poly(ethylene glycol) as a reinforcement for poly(lactic acid) nanocomposite scaffolds for potential tissue engineering applications. *Journal of the Mechanical Behavior of Biomedical Materials* 2016; 53:403-13.
[21] Wu D, Samanta A, Srivastava RK, Hakkarainen M. Nano-Graphene Oxide Functionalized Bioactive Poly(lactic acid) and Poly(ε-caprolactone) Nanofibrous Scaffolds. *Materials* 2018.

[22] Chiesa E, Dorati R, Pisani S, Bruni G, Rizzi LG, Conti B, et al. Graphene Nanoplatelets for the Development of Reinforced PLA–PCL Electrospun Fibers as the Next-Generation of Biomedical Mats. *Polymers* 2020.

[23] Zhao X, Luo J, Fang C, Xiong J. Investigation of polylactide/poly(ε-caprolactone)/multi-walled carbon nanotubes electrospun nanofibers with surface texture. *RSC Advances* 2015; 5:99179-87.

[24] Liu S, Wu G, Chen X, Zhang X, Yu J, Liu M, et al. Degradation Behavior In Vitro of Carbon Nanotubes (CNTs)/Poly(lactic acid) (PLA) Composite Suture. *Polymers* 2019.

[25] Bortoli LSD, Farias Rd, Mezalira DZ, Schabbach LM, Fredel MC. Functionalized carbon nanotubes for 3D-printed PLA-nanocomposites: Effects on thermal and mechanical properties. *Materials Today Communications* 2022; 31:103402.

[26] Mai F, Habibi Y, Raquez J-M, Dubois P, Feller J-F, Peijs T, et al. Poly(lactic acid)/carbon nanotube nanocomposites with integrated degradation sensing. *Polymer* 2013; 54:6818-23.

[27] Pötschke P, Andres T, Villmow T, Pegel S, Brünig H, Kobashi K, et al. Liquid sensing properties of fibres prepared by melt spinning from poly(lactic acid) containing multi-walled carbon nanotubes. *Composites Science and Technology* 2010; 70:343-9.

[28] Kobashi K, Villmow T, Andres T, Pötschke P. Liquid sensing of melt-processed poly(lactic acid)/multi-walled carbon nanotube composite films. *Sensors and Actuators B: Chemical* 2008; 134:787-95.

[29] Chakraborty G, Pugazhenthi G, Katiyar V. Exfoliated graphene-dispersed poly (lactic acid)-based nanocomposite sensors for ethanol detection. *Polymer Bulletin* 2019; 76:2367-86.

[30] Kumar B, Castro M, Feller JF. Poly(lactic acid)-multi-wall carbon nanotube conductive biopolymer nanocomposite vapour sensors. *Sensors and Actuators B: Chemical* 2012; 161:621-8.

[31] Chakraborty G, Katiyar V, Pugazhenthi G. Improvisation of polylactic acid (PLA)/exfoliated graphene (GR) nanocomposite for detection of metal ions (Cu^{2+}). *Composites Science and Technology* 2021; 213:108877.

[32] Han HS, You J-M, Jeong H, Jeon S. Synthesis of graphene oxide grafted poly(lactic acid) with palladium nanoparticles and its application to serotonin sensing. *Applied Surface Science* 2013; 284:438-45.

[33] Mokhena TC, John MJ, Mochane MJ, Tsipa PC. Application of Electrospun Materials in Oil–Water Separations. *Electrospun Materials and Their Allied Applications* 2020:185-213.

[34] Mokhena TC, Jacobs V, Luyt AS. A review on electrospun bio-based polymers for water treatment. *Express Polym Lett* 2015; 9:839-80.

[35] Mokhena TC, John MJ. Cellulose nanomaterials: new generation materials for solving global issues. *Cellulose* 2020; 27:1149-94.

[36] Yuan H, Zhang M, Pan Y, Liu C, Shen C, Chen Q, et al. Microspheres Modified with Superhydrophobic Non-Woven Fabric with High-Efficiency Oil–Water Separation: Controlled Water Content in PLA Solution. *Macromolecular Materials and Engineering* 2022; 307:2100919.

[37] Tuteja A, Choi W, Ma M, Mabry JM, Mazzella SA, Rutledge GC, et al. Designing Superoleophobic Surfaces. *Science* 2007; 318:1618-22.
[38] Wang X, Pan Y, Yuan H, Su M, Shao C, Liu C, et al. Simple fabrication of superhydrophobic PLA with honeycomb-like structures for high-efficiency oil-water separation. *Chinese Chemical Letters* 2020; 31:365-8.
[39] Wang X, Pan Y, Liu X, Liu H, Li N, Liu C, et al. Facile Fabrication of Superhydrophobic and Eco-Friendly Poly(lactic acid) Foam for Oil–Water Separation via Skin Peeling. *ACS Applied Materials & Interfaces* 2019; 11:14362-7.
[40] Yuan D, Zhang T, Guo Q, Qiu F, Yang D, Ou Z. Superhydrophobic Hierarchical Biomass Carbon Aerogel Assembled with TiO2 Nanorods for Selective Immiscible Oil/Water Mixture and Emulsion Separation. *Industrial & Engineering Chemistry Research* 2018; 57:14758-66.
[41] Yin Z, Pan Y, Bao M, Li Y. Superhydrophobic magnetic cotton fabricated under low carbonization temperature for effective oil/water separation. *Separation and Purification Technology 2021*; 266:118535.
[42] Sankaranarayanan S, Lakshmi DS, Vivekanandhan S, Ngamcharussrivichai C. Biocarbons as emerging and sustainable hydrophobic/oleophilic sorbent materials for oil/water separation. *Sustainable Materials and Technologies* 2021; 28:e00268.
[43] Saji VS. *Carbon nanostructure-based superhydrophobic surfaces and coatings.* 2021;10:518-71.
[44] Xing R, Yang B, Huang R, Qi W, Su R, Binks BP, et al. Three-Dimensionally Printed Bioinspired Superhydrophobic Packings for Oil-in-Water Emulsion Separation. *Langmuir* 2019; 35:12799-806.

About the Authors

Professor M. J. Mochane is teaching chemistry at undergraduate level and is a researcher in the Faculty of Health and Environmental Sciences at the Central University of Technology, Free State. He has published 37 articles (Reviews and research papers), 55 book chapters and attended different local and international conferences. Currently, he is an NRF Y-Rated researcher (National Research Foundation, South Africa), future professor programme (FFP) participant and (Professional Natural Scientist)), Pr. Sci.Nat and SACI. He is supervising masters and doctoral students in the Faculty of Health and Environmental Sciences (Department of Life Sciences). His research interest includes natural fiber reinforced composites, natural fiber hybrid polymer composites and Biopolymer/filler(s) composites for advanced applications.

Dr. Teboho Clement Mokhena obtained his PhD degree in Polymer Science at the University of the Free State, South Africa in 2017. After his PhD, he joined the University of Zululand as a senior Lecturer teaching physical chemistry from first year level to honors' level for 5 months. He, then, spent two years as researcher at Council for Scientific and Industrial Research (CSIR) under Nelson Mandela University (NMU). In June 2020, he joined Tshwane University of Technology (TUT) as researcher stationed at Council for Scientific and Industrial Research (CSIR) working on flame retardant of polymeric materials for three-dimensional (3D) printing applications. On the 1st of August 2021, he joined at Advanced Materials Division, MINTEK as a research scientist working on nanostructured materials for advanced applications.

Index

#

3D printing, 45, 46, 47, 49, 67, 68, 113, 131, 133, 138, 154

A

agricultural, 39
agriculture, 114, 124, 147
application(s), vii, 1, 2, 3, 11, 16, 17, 18, 19, 22, 33, 36, 37, 39, 40, 44, 45, 46, 48, 50, 54, 58, 66, 67, 68, 73, 87, 88, 89, 90, 91, 93, 94, 111, 112, 113, 114, 124, 140, 143, 144, 147, 148, 149, 151, 152, 153, 154, 155, 157, 158, 159, 160, 163
area, 2, 22, 45, 53, 57, 60, 67, 88, 121, 142
automotive, 2, 39, 93

B

barrier, 54, 59, 69, 74, 75, 80, 83, 84, 87, 96, 97, 100, 103, 107, 141, 149, 150, 159
biocomposite(s), vii, 37, 110, 144
biodegradability, 22, 39, 47, 73, 113, 148, 154, 157
biodegradable, 19, 39, 66, 67, 68, 69, 70, 93, 100, 110, 114, 140, 141, 147, 148, 158, 159
biomedical, 2, 39, 40, 54, 124, 141, 142, 144, 159, 160
bioplastics, 38, 93
biopolymer(s), vii, 19, 20, 37, 39, 66, 89, 93, 113, 140, 148, 156, 158, 160, 163
biotechnology, 147, 159
bottom-up, 1, 11

C

carbon, vii, 1, 2, 3, 4, 5, 6, 8, 9, 10, 11, 13, 14, 15, 16, 17, 21, 39, 40, 41, 43, 45, 46, 47, 48, 49, 50, 55, 57, 58, 59, 62, 63, 65, 66, 67, 68, 69, 70, 71, 73, 74, 75, 87, 88, 89, 90, 91, 93, 94, 96, 106, 108, 109, 112, 113, 114, 115, 116, 117, 122, 123, 126, 128, 132, 133, 136, 137, 138, 139, 141, 142, 143, 144, 145, 147, 148, 150, 151, 152, 154, 155, 157, 158, 159, 160, 161
carbon based fillers (CBFs), 1, 40, 42, 65, 74, 75, 81, 88, 93, 113, 116, 117, 122, 123, 124, 139, 154, 156, 157
carbon nanotubes, vii, 1, 5, 6, 8, 9, 11, 15, 16, 17, 40, 47, 66, 67, 68, 69, 70, 71, 89, 90, 91, 106, 108, 110, 112, 117, 141, 142, 143, 144, 145, 160
carbon-based fillers, vii, 1, 15, 39, 40, 73, 93, 94, 96, 109, 114, 127, 158
carbon-based nanomaterials (CBN), 1, 2, 3, 15, 53, 70, 159
carboxylic acid (COOH), 19, 33, 34, 35, 38, 41, 42, 43, 53, 57, 63, 64, 75, 76, 80, 117, 118, 120, 121, 132
catalyst, 5, 6, 8, 9, 13, 15, 17, 19, 25, 27, 29, 31, 33, 36, 63, 64, 71, 80, 107

Index

chemical exfoliation, 1
chemical interaction, 76, 113, 122
chemical vapour deposition (CVD), 1, 9, 11, 12, 13, 14, 15, 18, 107
composites, vii, 15, 16, 17, 39, 40, 41, 42, 43, 45, 46, 47, 48, 49, 50, 51, 52, 53, 54, 55, 56, 57, 59, 60, 62, 63, 65, 66, 67, 68, 69, 70, 71, 73, 74, 75, 76, 77, 78, 80, 81, 83, 84, 85, 87, 88, 89, 90, 91, 93, 97, 99, 103, 106, 108, 109, 110, 111, 112, 113, 114, 115, 116, 117, 120, 121, 123, 124, 125, 126, 128, 129, 130, 131, 135, 136, 137, 138, 139, 140, 141, 142, 143, 144, 145, 147, 148, 149, 150, 151, 152, 153, 154, 155, 156, 157, 158, 159, 160, 163
consumer goods, 19

D

degradation, 26, 31, 35, 57, 66, 67, 69, 73, 74, 79, 80, 83, 84, 86, 87, 89, 90, 91, 103, 104, 110, 111, 143, 147, 148, 154, 155, 160
dehydration, 25, 26, 36, 43, 116
drug release, 19

E

electronics, 124, 147
elongation-at-break, 53, 58, 115, 117, 118, 119, 120, 123, 126, 127, 128, 129, 130, 131, 132, 133, 134, 138, 139, 150
engineering, 16, 25, 37, 40, 68, 69, 70, 73, 90, 93, 110, 111, 112, 113, 142, 143, 148, 152, 153, 158, 159, 160, 161

F

fabricated, 5, 6, 9, 10, 20, 25, 27, 29, 100, 106, 108, 144, 161
filler(s), vii, 1, 15, 39, 40, 41, 42, 43, 44, 45, 48, 50, 51, 52, 53, 54, 56, 58, 59, 61, 62, 63, 64, 65, 66, 70, 73, 74, 75, 76, 78, 80, 82, 83, 85, 87, 88, 93, 94, 96, 97, 100, 102, 103, 107, 109, 110, 113, 114, 115, 116, 117, 118, 119, 120, 121, 122, 124, 126, 127, 128, 129, 130, 132, 135, 136, 137, 139, 142, 143, 144, 147, 148, 149, 150, 151, 155, 156, 158, 163
filler-modification, 113
flame resistance, vii, 93, 103, 105, 109
flame retardancy, vii, 93, 94, 95, 96, 101, 106, 109, 110, 111, 112, 140
flame-retardant, 93, 94, 96, 97, 100, 102, 103, 106, 108, 109, 110, 111, 112
flammability, 93, 94, 96, 97, 99, 103, 106, 108, 109, 110, 111, 112
fullerenes, 1, 5, 11, 17
functionalization, 33, 38, 41, 42, 47, 56, 57, 58, 71, 73, 80, 84, 118, 126, 131

G

graphene, 1, 5, 9, 10, 11, 12, 14, 15, 16, 17, 18, 40, 48, 49, 54, 59, 60, 61, 62, 64, 67, 68, 69, 70, 71, 73, 83, 84, 86, 87, 89, 90, 91, 101, 102, 103, 112, 117, 121, 126, 127, 129, 131, 133, 142, 143, 144, 150, 151, 152, 154, 157, 158, 159, 160
graphene oxide, 1, 5, 10, 11, 16, 17, 59, 61, 64, 69, 70, 71, 84, 86, 89, 90, 101, 102, 103, 112, 122, 142, 143, 150, 151, 152, 158, 159, 160
graphite, vii, 1, 5, 9, 10, 12, 14, 15, 16, 17, 18, 64, 83, 89, 90, 91, 96, 97, 99, 102, 110, 111, 140, 144

H

halogen free-flame retardant fillers, 93
health, vii, 114, 147, 163
heat, 15, 73, 74, 75, 82, 83, 96, 97, 100, 104, 107, 109, 110, 141
high-temperature applications, 73

Index

hybrid, 15, 16, 45, 78, 88, 89, 91, 110, 136, 137, 138, 139, 140, 143, 144, 145, 151, 154, 159, 163

hybridization, 49, 73, 88, 89, 119, 136, 137, 140, 151

I

impact strength, 113, 114, 118, 119, 124, 130, 132, 133, 134, 137

implants, 19, 113, 154

interaction, 16, 41, 42, 43, 44, 45, 47, 49, 50, 51, 53, 54, 57, 59, 61, 63, 64, 65, 74, 75, 76, 77, 80, 82, 83, 85, 87, 88, 107, 115, 116, 117, 118, 120, 121, 122, 124, 126, 128, 130, 132, 133, 134, 135, 137, 138, 139, 148, 154

isomeric forms, 19, 25, 115

L

lactide, 19, 23, 24, 25, 26, 29, 30, 31, 33, 36, 37, 38, 62, 63, 64, 67, 68, 69, 70, 71, 82, 85, 90, 91, 115, 120, 122, 141, 142, 144

M

mask, 147, 153, 154

mechanical exfoliation, 1

mechanical properties, vii, 15, 26, 41, 43, 44, 46, 53, 54, 57, 67, 68, 73, 90, 91, 110, 111, 113, 114, 115, 116, 117, 118, 119, 121, 123, 124, 125, 128, 130, 131, 132, 133, 135, 136, 137, 138, 139, 140, 141, 142, 143, 144, 149, 152, 160

medicinal applications, 19

medicine, 147

melt-blending, 39, 40, 41, 42, 46, 49, 108

melt-spinning, 39, 41, 51, 67, 68, 135, 154

modification, 19, 26, 31, 36, 38, 40, 41, 42, 43, 47, 54, 55, 56, 58, 60, 61, 62, 64, 65, 75, 78, 80, 82, 83, 85, 89, 103, 106, 110, 111, 113, 115, 116, 117, 118, 119, 121, 126, 128, 130, 131, 132, 134, 135, 139, 150, 154, 158

morphology, 5, 6, 9, 39, 40, 43, 57, 66, 70, 90, 115, 117, 152

O

oil-water separation, vii, 147, 157, 161

P

packaging, vii, 19, 39, 40, 73, 91, 113, 114, 116, 141, 147, 148, 149, 150, 151, 158, 159

PLA matrix, vii, 19, 31, 34, 47, 50, 57, 60, 63, 73, 93, 94, 96, 99, 100, 102, 103, 126, 148, 158

PLA/carbon-based fillers, 73, 93, 110, 113, 147

PLA/CBF composites, 39, 51, 65, 115, 116, 118, 125, 132, 139

polylactic acid (PLA), vii, 19, 20, 21, 22, 25, 26, 28, 29, 30, 31, 33, 34, 35, 36, 37, 38, 39, 40, 41, 42, 43, 44, 45, 46, 47, 48, 49, 50, 51, 52, 53, 54, 55, 56, 57, 59, 60, 61, 62, 63, 64, 65, 66, 67, 68, 69, 70, 73, 74, 75, 76, 77, 78, 79, 80, 81, 82, 83, 84, 85, 86, 87, 88, 89, 90, 91, 93, 94, 95, 96, 97, 99, 102, 103, 104, 105, 106, 108, 109, 110, 111, 112, 113, 114, 115, 116, 117, 118, 119, 120, 121, 122, 123, 124, 125, 127, 128, 129, 130, 131, 132, 133, 134, 135, 136, 137, 138, 139, 140, 141, 142, 143, 144, 145, 147, 148, 149, 150, 151, 152, 153, 154, 155, 156, 157, 158, 159, 160, 161

polymerization, 19, 20, 25, 26, 29, 30, 31, 32, 33, 36, 38, 40, 42, 58, 62, 63, 64, 65, 70, 71, 78, 82, 90, 116, 120, 123, 126, 130

preparation, 16, 37, 39, 40, 41, 42, 45, 46, 47, 49, 51, 53, 55, 56, 57, 59, 62, 63, 65, 66, 67, 69, 71, 75, 76, 78, 81, 89, 90, 91, 111, 112, 114, 116, 123, 125, 136, 139, 142, 144, 145, 156
protective char, 93, 104, 110
pyrolysis, 1, 9, 12, 96, 107

Q

quantum dots, 1

R

renewability, 39, 47, 113, 147
renewable resources, vii, 20, 93
resistance, 93, 94, 96, 103, 109, 110, 114, 156

S

sensing, vii, 67, 68, 89, 147, 155, 156, 158, 160

solution mixing, 39, 40, 42, 52, 59, 65, 81, 101, 117, 120, 156
stability, 14, 54, 67, 69, 73, 74, 75, 76, 80, 83, 84, 87, 88, 89, 90, 91, 93, 111, 143
sustainability, 39, 73, 147, 157, 158
synthesis, vii, 1, 3, 5, 6, 9, 11, 12, 14, 15, 16, 17, 18, 19, 20, 21, 22, 25, 26, 28, 30, 31, 32, 33, 36, 37, 38, 71, 80, 90, 112, 147, 159, 160

T

tensile strength, 44, 53, 58, 113, 114, 117, 118, 119, 120, 122, 123, 124, 125, 126, 127, 128, 129, 130, 131, 132, 133, 134, 135, 137, 138, 139, 150, 154
textile, 39, 40, 111
thermal stability, vii, 16, 40, 54, 66, 68, 69, 73, 74, 75, 76, 77, 78, 79, 80, 81, 82, 83, 84, 86, 87, 88, 90, 111, 112, 116, 141, 148, 151, 158
top-up, 1, 11